The Bone Hunters

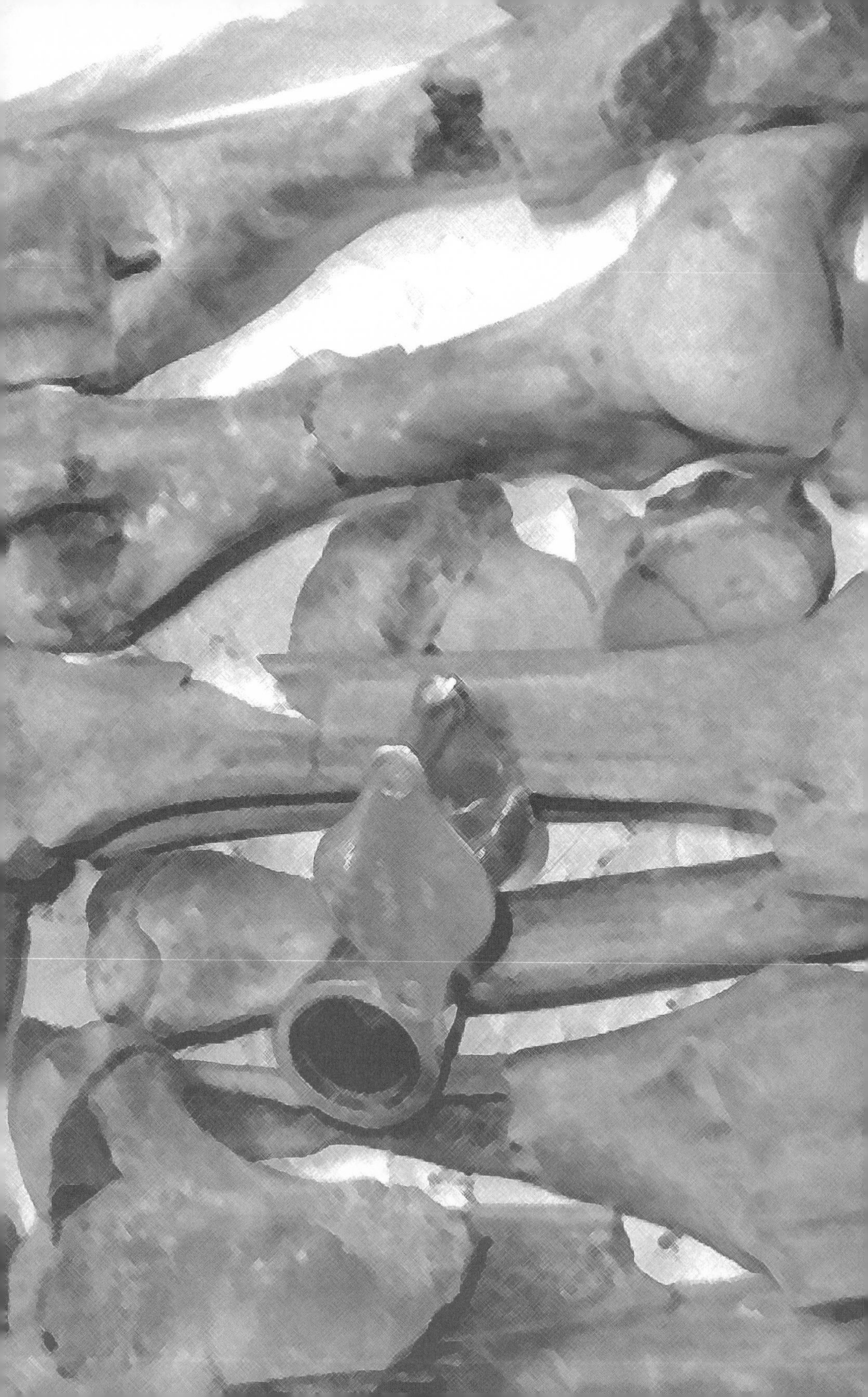

The Bone Hunters

The Discovery of Miocene Fossils in Gray, Tennessee

Harry Moore

The University of Tennessee Press

Knoxville

First Edition.

Unless otherwise credited, all photographs are by the author.

This book is printed on acid-free paper.

Library of Congress Cataloging-in-Publication Data

Moore, Harry, 1949–
The bone hunters: the discovery of Miocene fossils in
Gray, Tennessee / Harry Moore.— 1st ed.
p. cm.
Includes bibliographical references and index.
ISBN 1-57233-323-5 (pbk.: acid-free paper)

1. Vertebrates, Fossil—Tennessee—Gray.
2. Paleontology—Miocene.
I. Title.

QE841.M74 2004
566'.09768'97—dc22 2004005631

The important thing is not to stop questioning. Curiosity has its own reason for existing. One cannot help but be in awe when he contemplates the mysteries of eternity, of life, of the marvelous structure of reality. It is enough if one tries merely to comprehend a little of this mystery every day.

—Albert Einstein

Contents

Illustrations

Figures

Maps

Table

Acknowledgments

In writing this book about the discovery and saving of the Gray Fossil Site, I found that I am grateful to others on two levels. On one level, I wish to express my personal thanks to all those who assisted me in the development of this work. On another level, I want to express my deep appreciation—which should be shared by anyone who loves Tennessee and values her scientific treasures—to the many individuals who worked at the site, grasped its significance, and endeavored to rescue it for posterity. To a large extent, these two groups of people overlap.

Larry Bolt deserves a special nod. His early insight that the layered clay deposits being uncovered at the Gray road project might be worthy of closer inspection did much to set the salvage operation in motion. His dedication, perseverance, and excitement helped me and others to recognize the importance of the find. It was also Larry, along with Martin Kohl, who found the first fossil bone pieces at the site.

Without the assistance of Marta Adams and her husband, Tony Underwood, and of Paul Parmalee and Walter Klippel of the University of Tennessee, the larger meaning of the fossil discovery might never have been realized. The long, unselfish hours they spent in excavating fossil bones—followed by equally long hours devoted to cleanup, preservation, and identification—were invaluable and are most graciously acknowledged.

Bob Price and Peter Lemiszki of the Tennessee Division of Geology deserve thanks for their interest and initial exploration of the site, which aided in understanding the deposit. Special thanks should also go to Mike

Clark, Bob Hatcher, Chris Wisner, and Don Byerly of the University of Tennessee Department of Geological Sciences and Jeff Munsey of the Tennessee Valley Authority for their expert field interpretations of the geological aspects of the fossil site.

Nick Fielder merits recognition for his help and leadership during the circumstances that unfolded after the initial announcement of the fossil site find. Further appreciation goes to Lanny Eggers and Freddie Holly, who were deeply involved in the actual discovery of the deposit and the contractual activities that took place there afterwards.

Special thanks and recognition are owed to Larry Hathaway for his untiring dedication in finding and salvaging a plethora of fossil bones from the construction work. Rick Noseworthy and Keven Brown also deserve deep appreciation for their help and commitment in salvaging fossil material, especially the alligator skull that Rick found. In addition, thanks should go to Rab Summers of Summers Taylor, Inc., for taking measures to ensure that his construction personnel aided our efforts in locating and rescuing delicate fossils. I must also acknowledge and thank Luanne Grandinetti for her superior handling of media relations during the events surrounding the discoveries at Gray.

I am most grateful to a very supportive staff at TDOT's Geotechnical Engineering office in Knoxville, including Nancy Chadwell, George Danker, Saieb Haddad, David Barker, Carl Elmore, Conley Lynn, Teddy Plemmons, Ledford Russell, and Jimmy McGill. Demonstrating both patience and professionalism, Andrea Hall, TDOT roadway designer, worked with me to incorporate fossil-site details into the roadway project plans; she was also responsible for the road relocation design. In addition, I thank Vanessa Bateman for her enthusiasm and willingness to work long hours to extract some of the fossil bones. Len Oliver, my section manager, merits thanks for the ready support he gave to such an unusual undertaking on a roadway project.

Without the understanding and backing of Don Sundquist, former governor of Tennessee, and Bruce Saltsman, former transportation commissioner, the saving of this fossil deposit would probably not have been possible. Therefore, considerable appreciation is extended to them and their staffs.

It should go without saying that minus the distinctive contributions of all of the above people to the preservation of the Gray Fossil Site, this book would never have happened.

In addition, special thanks are given to Gene Adair for his expert editorial advice and his ability to guide me in the clarification of a most unusual subject. I am grateful as well to Jennifer Siler and the staff of the University of Tennessee Press for their willingness to support yet another book on the geology of Tennessee. And for his encouragement and advocacy of this project, I thank Scot Danforth, UT Press editor.

Finally, I am most thankful to my wife, Alice Ann, for her never-ending support of my geology book endeavors.

Prologue
Creatures Emerging from the Clay

"Quick, Vanessa, hand me the small trowel," I said. "I think I can remove the clay from this part of the bone while you spray water on it."

On this hot day in June 2000, my colleague Vanessa Bateman and I were digging feverishly in the dark gray clay of a freshly excavated roadbed in northeast Tennessee. Our tools had hit some hard objects, and we soon discovered that they were not rocks. They were fossilized bones, and the one I found included a row of teeth. Perhaps it was an animal jawbone, I thought. Meanwhile, Vanessa's diggings a few feet away revealed a row of vertebrae with rib bones jutting out on either side.

Turning to help me, Vanessa used our plastic water bottle to wash the mud from the teeth exposed in the bone I had uncovered. "I can't believe how sticky this clay can be," she said as she squeezed water over the rusty brown fossil while I scraped away at its edges. What I had thought at first to be only a jawbone turned out to be a skull from some sort of animal, and it was largely intact.

Being rather husky and not very tall, I was able to curl up easily beside the hole we had excavated and remove various rocks, some the size of watermelons, that were mixed in among the clay and bones. Although I routinely worked out with weights and walked two and a half miles a day, I was wearing out quickly from the heat and the exertion of digging. But the excitement of extracting these ancient bones from the earth kept me going.

Vanessa's own infectious enthusiasm helped a lot. Like me, she was a geologist employed by the Tennessee Department of Transportation (TDOT). Young and energetic, she had graduated with a bachelor's degree in geology from Middle Tennessee State University a few years earlier. She had joined the TDOT Geotechnical Engineering office in Nashville from the private engineering sector, where she had worked for a year or two coming straight out of college. She was full of questions and ideas, and I enjoyed being around her.

As we continued our efforts, Vanessa used a knife to scrape the clay away from one side of the skull, while I worked the other side with a trowel. Our immediate goal in uncovering the skull and nearby vertebrae was to box them up for transfer to the University of Tennessee at Knoxville, where they could be identified and preserved. Only a few weeks before, the road project had exposed the fresh clay and numerous bones. The site had already attracted the curious, and we were determined to salvage as many bones as possible before souvenir hunters ravaged the excavation.

"I'll wash the clay from the bone with the fluff brush, and then we can lift the skull together," I told Vanessa. We could not believe that we had unearthed the skull and backbone of a prehistoric animal right here in Tennessee. Later identification by University of Tennessee experts revealed that what we had just uncovered were parts of a fossil tapir more than five million years old.

Reconstructive efforts by paleontologists since the discovery of fossil bones near Gray, Tennessee, in May 2000 have uncovered important clues to a prehistoric Tennessee landscape that is as intriguing as it is unbelievable. Thanks to the discoveries at what is now called the Gray Fossil Site, one can vividly imagine what the area may have looked like millions of years ago—a time when the landscape of East Tennessee was not unlike that of the modern Serengeti Plain in Africa.

■ ■ ■

A breeze was blowing and clouds were forming in the sky—those white, fluffy, cotton-like clouds that appear after a weather front has moved through. The summer air was hot, clear, and fresh. Insects buzzed around animals and just above the surface of a pond. Some of the insects clumped together in swarms as they glided over the water. The continuous cadences of crickets filled the air and mixed with frog calls. Shallow in most places, the pond water was murky where animals had trotted out for a drink, stirring up mud and refreshing themselves with a quick dousing.

A frog poked out of the water to rip insects from the air. Croaks could be heard all around as other amphibians sounded in the humid fragrance of the pond. An alligator lay in wait for its next meal along the shallows, beneath overhanging vegetation. Another alligator floated in the middle of the pond with its eyes protruding just above the surface. Ripples formed around the snout as it breathed deeply before going under.

The water of the lake teemed with life. Turtles sunned themselves on old tree snags and floating logs, while snakes lay among the water grasses. An occasional ripple would break the calm of the surface as a sunfish flipped its tail fin and dove back into deeper water.

Tapirs, mammals that looked like a cross between a hog and an anteater, waded nonchalantly in the water's edge, browsing on the abundant plants that surrounded the watering hole. Infant tapirs, the size of small dogs, huddled with their mother while adults foraged on the abundant vegetation around the pond. Their snorting and sloshing stirred up numerous flies and other insects.

Nearby, an elephant spewed water from its trunk to wash the dust and dirt from its back. The tapirs scattered from the noise. The elephant sounded its bugle to warn the others of his dominance. Rhinoceroses also fed on the grasses and shrubs that lined the pond. Short, stocky creatures with low bellies that hung a few inches above the ground, the rhinos were colored a grayish brown from the mud that had caked on their leathery, gray skin.

Some six hundred feet in diameter, the pond provided drinking water for all the animals and sustenance for the surrounding vegetation. A towering pinnacle of gray limestone bounded one side of this watering hole; at the base of this cliff was a cavern from which a constant flow of spring water fed the pond.

Vines draped the opening of the cavern and filtered the daylight before it melted into the darkness of the subterranean depths. The grayish brown limestone faded into the abyss of the subsurface where the resurgence from the cave stream bubbled into the surface pond. Damp, cool, musty air issued out of the cave entrance and blew against the vines and tree branches that hung over it. An occasional bird inspected the limestone bluff overlooking the pond.

It was, all in all, a bucolic scene—but one where occasional mishaps also occurred. The water was deepest next to the limestone bluff, and any animal venturing out that far was likely to drown. And when that happened, its bones would eventually become sealed in the muddy sediments at the bottom of the pond. As the millennia passed and the terrain underwent countless changes, the bones gradually fossilized. And finally, during the middle of A.D. 2000, they would be exposed to the air once more.

Introduction
Geology and Fossils in Tennessee

This book is about the discovery of a major fossil deposit near the town of Gray in upper East Tennessee and the events that followed that discovery. It is a story that involves not only the fossil bones but also the people of the nearby community and the ways in which construction of a highway played a major role in the find. To set the stage for my recounting of these events, it may prove helpful to review some aspects of geology, particularly that of the East Tennessee area, and to put into context the significance of the Gray Fossil Site.

Readers less interested in the finer details of the science may want to skip ahead to chapter 1, where my narrative of the Gray Fossil Site discovery and salvage operation actually begins. I hope, however, that most readers will devote some attention to the next few pages, for I believe that a basic understanding of Tennessee geology and the processes of fossilization can only enhance their appreciation of the dramatic story that unfolded at the Gray Site.

Beginning in the late 1800s, the early geologists of Tennessee had a mission to find and catalogue the rock strata of the state. They looked in the river bottoms, on the hills, and atop the mountains. They mainly found layers of sandstone, shale, limestone, slate, and quartzite, which are all rocks of sedimentary origin. Much less commonly, and largely in the extreme eastern portion of the state, they found granite, a rock of igneous origin—that is, one formed when molten rock within the earth pushes to the surface and solidifies. In time, the collective knowledge of Tennessee geology became extensive and was described and summarized in geologic maps and detailed written reports.

Geologists and geographers have resolved the landscape across Tennessee into a number of landforms known as physiographic provinces. These areas of like surface features are divided, from east to west, into the following: Blue Ridge, Valley and Ridge, Cumberland Plateau, Eastern Highland Rim, Central Basin, Western Highland Rim, Western Valley of the Tennessee River, Gulf Coastal Plain, and Mississippi River Alluvial Flood Plain. In addition to surface features, the geologic strata of these provinces have been defined and catalogued.

Although each of the Tennessee's physiographic provinces have special and distinct features involving topography, geology, plant and animal diversity, and human use patterns, only the three easternmost provinces will be further discussed here because they are the most relevant to the subject of this book. And of these, the Valley and Ridge province is the most significant, since that is where the Gray Fossil Site was found.

The Blue Ridge province, which lies immediately east of the Valley and Ridge, consists of the high mountain regions of extreme eastern Tennessee,

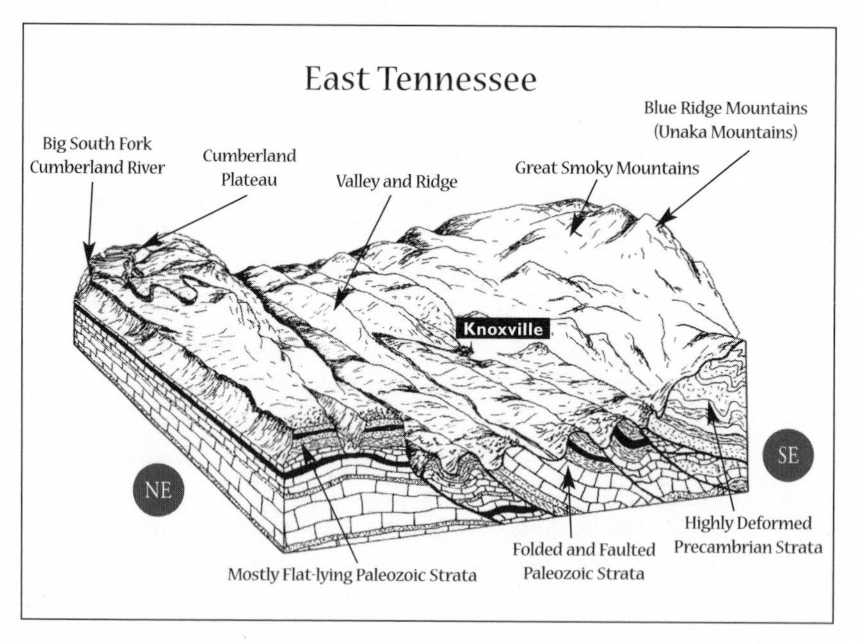

A schematic drawing illustrating the general geology of East Tennessee in relation to the physiographic provinces. The Gray Fossil Site is located in the Valley and Ridge province.

with elevations ranging to over 6,000 feet; it includes the Cherokee National Forest and the Great Smoky Mountains National Park. Underlying the terrain are Precambrian-age rocks dating from 600 million to over 1 billion years ago; these include metamorphosed sedimentary rocks and some igneous and unaltered sedimentary strata. In some places (largely along the westernmost front of the province), younger Paleozoic-age rocks are exposed, usually in isolated coves such as Cades Cove and Tuckaleechee Cove.

The Cumberland Plateau, which borders the Valley and Ridge on the west, is generally a high, tabletop-like landscape that includes numerous waterfalls and deeply incised canyons. The plateau is bounded on the west and east by escarpments that range up to about 1,000 feet or more in height. The plateau also varies in elevation from about 2,000 to 3,000 feet and is underlain primarily by flat-lying Paleozoic-age rocks, including sandstone, siltstone, shale, and coal, which are 240 to 360 million years old.

Sandwiched between the Blue Ridge and the Cumberland Plateau, the Valley and Ridge province incorporates parallel trending ridges and valleys underlain by folded and faulted sedimentary rock of Paleozoic age. This region has been timbered and cultivated and, most recently, urbanized in several areas, making great expanses of natural vegetation somewhat uncommon. The major population centers of East Tennessee—Kingsport, Bristol, Johnson City, Greeneville, Morristown, Knoxville, Maryville, Sweetwater, Cleveland, and Chattanooga—are located in this province.

The folded and faulted rock of the Valley and Ridge tends to run parallel to the corresponding ridges and valleys. This is mostly due to the differential weathering of the sedimentary rock layers. The harder rocks, such as sandstone and some types of limestone and dolostone, are more resistant to weathering and tend to form the ridges, while the less resistant rocks, such as shale and shaley limestone, tend to form the valley floors.

It is also in the Valley and Ridge that large areas of caves and sinkholes have formed, often resulting in a rocky limestone landscape devoid of surface streams. Landscapes of this type are usually referred to as karst and karst terrane. The karst terrane, which includes such subsurface features as caverns and associated groundwater, is often the scene of collapse sinkholes as well as flooding along highways and in commercial and residential developments.

The subject of this book, the Gray Fossil Site, is found in the northernmost area of East Tennessee within the Valley and Ridge province. Folded and faulted Paleozoic Age sedimentary rocks, mainly of Cambrian

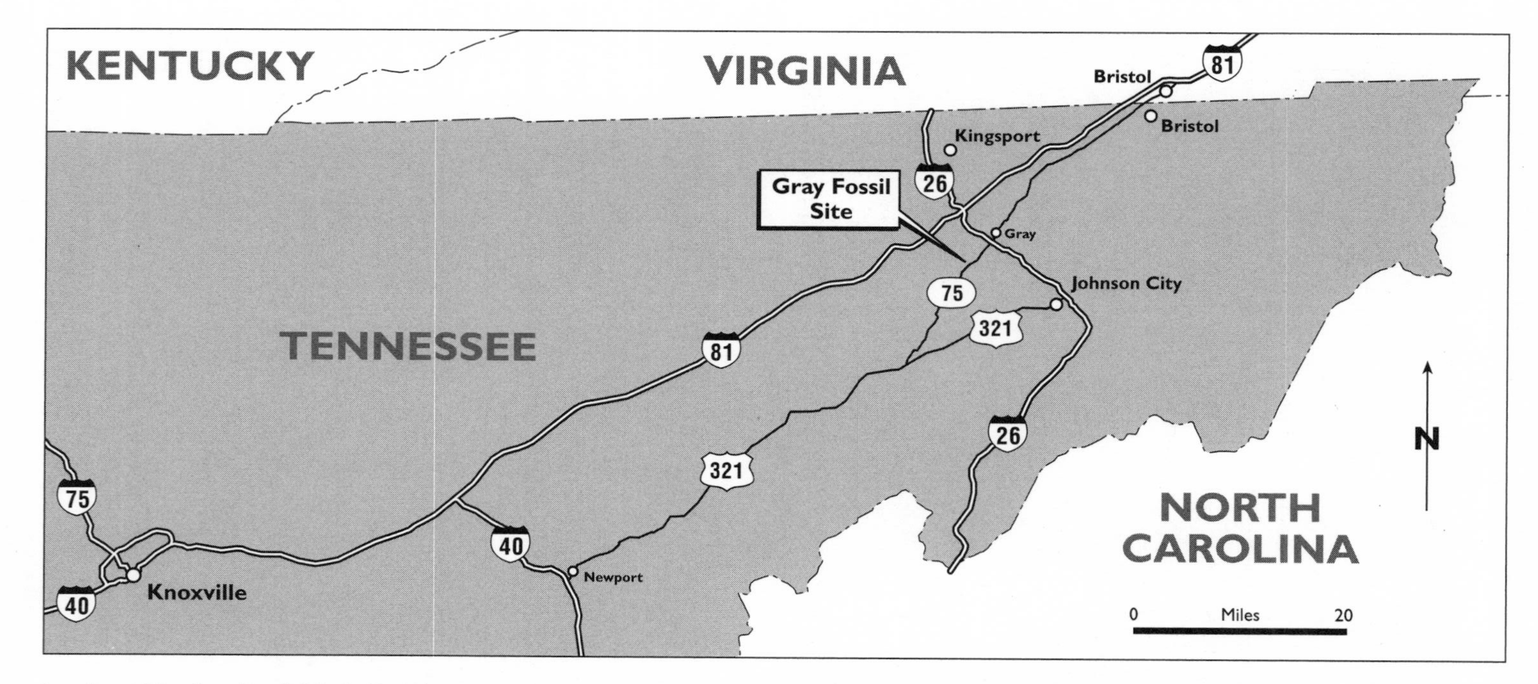

Location of the Gray Fossil Site in East Tennessee. At the time of the fossil discoveries, Interstate 26 was named Interstate 181.

An aerial view of the Gray Fossil Site just after its discovery in June 2000. Daniel Boone High School is visible in the upper left corner. TDOT photo by George Hornal.

and Ordovician age (430 million to 570 million years old), underlie this area. These rocks were folded by the crunching action of the colliding continents of North America and Africa some 230 million years ago, an event known as the Alleghanian Orogeny, or mountain-building episode. These rocks lie in long belts of strata trending from northeast to southwest, and because of weathering, they form long linear ridges and valleys that also trend from northeast to southwest. Most of these rocks are limestone, dolostone, sandstone, or shale.

The sedimentary origins of most of the rocks found in Tennessee make the possibility of uncovering the fossil remains of once-living organisms buried in those rocks very high. In fact, fossils are commonly found in the rocks all across the state, with the exception of the Blue Ridge province.

"Fossil" is a term that applies to plant or animal remains that have been preserved by natural causes in the earth's crust. Nature can preserve such remains in many ways, though not all of them occur in Tennessee. Organisms might undergo quick burial in sediments as a result of a landslide or by settling to the bottom of a body of water. Some might become encased in ice as a result of glacial movement. Remains may also be preserved by mummification, which occurs mainly in dry climates, or by entrapment in tar pits or amber resin. Fossilization often involves mineral replacement, in which the original tissue is dissolved and gradually replaced by the minerals that are soaked into its pores. Another process, carbonization, occurs when decaying tissues leave behind traces of carbon; plant remains that have been reduced to lignite, a form of coal, are examples of this process. Still other fossils are formed as cast-molds: in this process, an organism leaves behind an impression of itself in the surrounding sedimentary rock, and this hollow space is later filled by water containing dissolved minerals that eventually form a copy in stone of the original organism.

Most of the fossils found in Tennessee are of the mineral replacement, carbonization, and cast-mold types resulting from quick burial in sediments. Fossils in Tennessee rocks can be easily found in shale and shaley limestone, as these rocks typically form from "muddy" sediments that are more apt to seal and preserve the animal remains.

The most common fossils found in Tennessee rocks are the shells of brachiopods (marine shellfish), the reefs of bryozoa ("moss animals") and coral, and the stems of crinoids ("sea lilies"), which all date to the Paleozoic Age, or 230 to 570 million years ago. Other common fossils include the 230-million-year-old plant fossils of the Plateau region, mainly ferns and bulrushes, and the more recent, 30- to 50-million-year-old leaf impressions of the West Tennessee clay deposits near the towns of Camden and Paris.

One noteworthy deposit of fossil animals, known as the Coon Creek Fossil Site, is found near the community of Leapwood in McNairy County in West Tennessee. In the early 1900s, Bruce Wade of the United States Geological Survey found over three hundred species of invertebrate animals, mostly mollusks such as clams and snails, of Cretaceous age (approximately 65 million years old), along the banks of Coon Creek in the early 1900s. These shells were buried and beautifully preserved in micaceous gray clay marl that sealed them from the deteriorating effects of oxygen. As a result, some of these fossils still contain the "mother-of-pearl" layer found on the inside of the shells. In 1998 the Tennessee legislature named one of the Coon Creek fossils, that of a clam known as *Pterotrigonia,* as the State Fossil.

In East Tennessee, there is a large gap in the geologic record that spans from the end of the Paleozoic Era, some 230 million years ago, to the Pleistocene Epoch, commonly referred to as the "Ice Age," which began about

Geologic Time

In our modern world, we often seem obsessed with time. We think about time to go to work or school, time to go to bed, or time to do an important project. We often say to ourselves, "If only I had more time . . ." Time is also crucial to how geologists seek to understand the earth's makeup. However, geologic time is not merely a matter of seconds, minutes, hours, days, weeks, months, years, decades, or centuries. Rather, it encompasses millions and even billions of years.

Scientists tell us that the earth is approximately 4.5 billion years old—or 1 million years multiplied by 4,500. What we call civilization has been around for less than 10,000 years or 1 percent of a million years. The United States has been a nation for only 7.8 percent of a million days. For human beings with an average life span of 75 to 80 years—an infinitesimal "blip" in geologic time—such numbers strain the imagination. It becomes readily obvious that any block of time associated with humans and recorded human history is insignificant when compared to the many millions or hundreds of millions of years it takes for significant geological changes to occur.

Geologists divide time into many subdivisions. The largest and earliest block of time is called Precambrian, and it extended from the time of the earth's formation to about 570 million years ago—a period of about 4 billion years. The next largest subdivision, the Phanerozoic Eon, extends from the end of Precambrian time up through the present. As the accompanying table shows, the Phanerozoic Eon is further broken down into eras, periods, and epochs. And as the table also illustrates, various life forms came—and often went—during these particular blocks of time. The famous "Age of the Dinosaurs," for example, occurred during the Mesozoic Era and the three periods—Triassic, Jurassic, and Cretaceous—that constitute it. It lasted approximately 175 million years and ended some 65 million years ago.

In this book, two epochs from the "Age of Mammals," which followed the extinction of the dinosaurs, are repeatedly mentioned. The more recent epoch is the Pleistocene, commonly called the "Ice Age." Characterized by glaciation over much of the globe, it lasted more than 1.7 million years and ended only 10,000 years ago. The other much-referenced epoch (and one that is separated from the Pleistocene by the Pliocene Epoch) is the Miocene Epoch. Ending about 5 million years ago, it was a time when a warmer climate prevailed over much of the planet. As the reader will discover, scientific knowledge of the vertebrate life forms that existed during these epochs played a key role in determining the geologic age of the Gray Fossil Site. ■

1.8 million years ago and ended about 10,000 to 12,000 years ago. Weathering and erosion have largely removed any geologic deposits that may have been deposited in this region during that time span.

Time Units of the Geological Time Scale					Development of Plants and Animals
Eon	Era	Period	Epoch	Boundary	
Phanerozoic	Cenozoic	Quaternary	Holocene	0.01	Humans develop
			Pleistocene	2	
		Tertiary	Pliocene	5	
			Miocene	24	"Age of Mammals"
			Oligocene	36	
			Eocene	58	
			Paleocene	65	Extinction of dinosaurs and many other species
	Mesozoic	Cretaceous		140	First flowering plants
		Jurassic		205	First birds Dinosaurs dominant
		Triassic		240	Extinction of trilobites and many other marine animals
	Paleozoic	Permian		290	
		Pennsylvanian		330	Large coal swamps First reptiles
		Mississippian		360	Amphibians abundant
		Devonian		410	First insect fossils Fishes dominant
		Silurian		435	First land plants
		Ordovician		500	First fishes
		Cambrian		570	Trilobites dominant First organisms with shells
Proterozoic		Collectively called Precambrian, comprises over 85 percent of the geologic time scale			First multicelled organisms First one-celled organisms
Archean					Age of the oldest rocks
Hadean		4600			Origin of earth

Generalized geologic time scale. Adapted from the American Geological Institute (age represented in millions of years).

Colluvial boulder deposits, usually found along major bluffs and escarpments, and old river terrace deposits are the most common geologic evidence representing the last million years or so. In addition, the remains of "Ice Age" wooly mammoths and mastodons (both elephant-like animals), as well as those of tapirs and saber-tooth tigers, are occasionally found in old sinkholes and some cave deposits in East Tennessee. These tend to be the only vertebrate terrestrial fossils found in Tennessee.

At least that was the case until the discovery of the Gray Fossil Site.

Sometimes, just when we think we know all there is to know about a subject, we get a big surprise. That is what happened to the world of Tennessee geology in the late spring and summer of 2000. The discoveries that emerged during those months added a whole new chapter to our collective knowledge and make for a story that is still unfolding.

1 The Beginning

The discovery of the Gray Fossil Site began as an uncomplicated road construction project in the spring of 2000, one of dozens that the Tennessee Department of Transportation undertakes each year. This particular project was designed to realign the centerline of an approximately one-mile portion of State Route 75 (S.R. 75) in Washington County near the small community of Gray (population 1,071). The existing highway had a bad history of automobile accidents, and like most TDOT construction projects, this one was intended to make the roadway safer and more durable.

The area targeted for improvement lay about two miles south of the interchange of Interstate 181 (now Interstate 26) and S.R. 75 near Gray and halfway between Johnson City ánd Kingsport, approximately seven miles from each. Daniel Boone High School on S.R. 75 was about one-third of a mile south of the site. The construction was to involve straightening the horizontal curves of the road section and flattening its hills and dips. The improved section of road was designed for two lanes of traffic, with widened shoulders and an added center turn lane in appropriate locations.

This part of S.R. 75 winds through a rolling valley of open grass farmland with a scattering of homes. Clusters of trees surround some homes and the highest ridge tops. Occasionally, the gray hulk of a weathered and tattered barn can be seen along the road, especially south of Daniel Boone High School. Years ago, before the land was put to agricultural use, it was covered mostly with trees and scrub brush. Today, continued growth and eventual urbanization are expected in the area as developers build residential subdivisions and other properties.

TDOT designed the road construction project and put it out for bids in late 1999. Summers Taylor, Inc., of Elizabethton, Tennessee, submitted the winning bid of about $1.6 million. The work was set to begin in mid-May 2000.

Wherever new roads are being planned and designed, the TDOT Geotechnical Engineering Section routinely conducts geotechnical surveys and investigations of the locations. The route along the section of S.R. 75 corridor was subjected to the usual drilling and soil-sampling procedures that are conducted during the design phase of road projects.

The results revealed nothing out of the ordinary. The auger-type drilling machinery used by TDOT penetrated moderately stiff clay soils down to a point below the proposed roadway grade elevation. As is typical in East Tennessee, the surface soils consisted mainly of oxidized residual clays that were orange-brown to reddish-brown in color. Nearer to the bedrock, the drilling turned up some gray clay, but again, this was hardly unusual. In this part of the state, gray clays tend to develop along the soil-bedrock interface where the bedrock is limestone, composed mainly of the mineral calcite, or dolostone, composed mainly of the mineral dolomite. The gray clays result from the weathering of the limestone and dolostone bedrock.

The TDOT investigators duly noted the presence of the contrasting layers of clay and added this information to the roadway plans in preparation for the construction project. The department finalized the design of the road improvement and purchased the needed right-of-way. It seemed that this project would be uneventful and ordinary, like most of the other projects that TDOT constructs yearly, whether they involve road widening, repaving, bridge renovation, or new road construction. Thus, no one knew at the time that the stage had just been set for what would become a most amazing discovery in the history of science in Tennessee.

Every spring, TDOT construction engineering field offices become a beehive of activity, and the spring of 2000 was no different. As the cold, rainy winter gives way to weather more conducive to road building, the offices become actively engaged in new projects. Personnel typically busy themselves with construction inspection, contract issues, monthly estimate calculation for payments to contractors, and construction field problems. The field problems can vary from project to project but may include such issues as adjusting driveway connections for property owners and dealing with erosion, siltation pollution, sinkhole formation, landslides, and, in some instances, the presence of soft, wet clay soils.

On May 15, Freddie Holly, who works in the TDOT construction field office of Lanny Eggers in Elizabethton, Tennessee, called me with the news that the contractor had uncovered a soft gray-to-blackish clay deposit on the S.R. 75 road-widening project near Gray. In his characteristically soft-spoken, well-mannered way, Freddie described the problem: the soft clay was miring up the grading equipment and thus causing concerns about the future stability of the subgrade, or "roadbed" of the highway.

Freddie's supervisor, Lanny Eggers, had asked him to contact me. Lanny and I had just completed several years of working together on a multimillion-dollar road project in the nearby Unicoi County mountains on I-26. We had developed a good working relationship. I respected Lanny's ability to organize his office and conduct the hard and tedious fieldwork required in road construction projects. He was dependable and handled his responsibilities well. I trusted him and especially liked his ability to command his construction personnel.

I manage the TDOT Geotechnical Engineering office in Knoxville, which oversees the geological and geotechnical issues of road construction within a twenty-four-county area known as TDOT Region One. Our work involves not only identifying the geologic conditions along proposed road projects but also making design and construction recommendations regarding soil and rock cut slopes and embankment construction. In addition, we investigate the foundation conditions for proposed bridges and retaining walls for road projects and make the appropriate foundation-support recommendations. We also investigate sinkhole and cave problems as well as landslides that occur along the highway system. Whenever the Design, Construction, and Maintenance Divisions of TDOT encounter difficulties involving the soil and rock makeup of construction and improvement sites, they call on us for help.

The soft clay soil encountered at the Gray site presented a problem because projects of this type often use a construction method called "cut and fill." When contractors "cut" out the soil and rock along a road alignment, as they were doing at Gray, they usually use the excavated material to build up the "fill" sections or embankments of the road. This type of road construction is the most common and generally the least expensive method. In most instances, the excavated material meets the specifications that TDOT has outlined for such use of soil and rock. But sometimes the soil is too muddy and soft or contains too much organic material to be used in the road, and so it is typically "wasted" or disposed of in an off-road site. This was the case at the Gray project, and that was why Lanny Eggers had asked Freddie Holly to call me.

As Freddie explained, the soil at the Gray site was giving the contractor, Summers Taylor, plenty of headaches. The crew was spending valuable construction time pulling heavy equipment across and out of the blackish-gray clay as they tried to cut through a small hill for the new roadbed. In places the clay was almost mud-like, and it was hindering the mobility of the company's earth-moving machines, called "pans." These are massive scoop-type machines with tires as tall as a person. In ordinary conditions, a pan pivots around its front tires and drags its large shovel-like midsection over the soft ground, scooping the earth up into the middle of the machine. Here, however, the pans were simply spinning their huge tires in the soft, wet clay. The

Dark gray and black in color, the soft clay at the S.R. 75 construction site caused excavation problems that led TDOT geologists to the fossil discoveries. The tire ruts left by the heavy grading equipment attest to the material's softness. Photo by Larry Bolt.

contractor had to use bulldozers to push the pans out of the clay and onto firmer soil. Clearly, this problem had to be corrected.

The soft clay was obviously unacceptable as a subgrade or "foundation" material at the Gray site because heavy vehicular traffic would quickly cause the roadbed to settle, breaking up the surface asphalt. When highway construction crews encounter such soil, a process called undercutting usually removes it. This is simply the excavation of the soft soil down to a point below the future roadbed so that an acceptable material, such as crushed stone or approved clay, can be used to replace it. Such measures are usually undertaken at the recommendation of a geologist or engineer.

Lanny and Freddie wanted me to visit the construction site and see the ground for myself so that I could make the appropriate recommendations. I speculated that undercutting the soft material and replacing it with a rock pad would probably fix the problem, but this was only a tentative recommendation. An on-site inspection was needed before a final decision could be made.

Since I was to be out of town for the next few days on a trip to New York State, I asked whether my visit to Gray could wait until my return. No, Freddie told me, the construction office needed help as soon as possible. Thus, I asked one of our office geologists, Larry Bolt, to investigate the site. Larry loved geology and often engaged me in theoretical discussions about the origins of life and the earth. We had a comfortable working relationship, and I had faith in his judgment, especially about problems involving odd occurrences of rocks and soils.

Larry visited the project site on May 17. He met with Freddie Holly, who showed him where the problem had occurred at the intersection of Fulkerson Road with S.R. 75. When the construction workers had cut through the hill there, they encountered the soft clay at depths of five to fifteen feet. Inspecting the excavation area, Larry identified very soft, thinly laminated clays that consisted of alternating layers of whitish-brown and tan clay-sand averaging over twenty layers per inch of sediment. To the naked eye, this soil appeared to be gray. Among these layers of sandy clay were layers of a black clay-like material containing ancient plant debris—a mixture of carbonized leaves, twigs, and pieces of wood, which is commonly referred to as lignite. The clay was sandier in some spots than others and even contained gravel in a few places.

I read Larry's report on May 25, following my return from New York, and he and I debated its implications over the next week. The clays, Larry

The thinly laminated clay was unusual for the geology of the area. Such material typically results from sedimentary deposits in a still lake or large pond.

wrote, appeared to be varve-like. The term "varves" refers to contrasting layers of sediment representing seasonal sedimentation, which usually occurs in climates and environments where glaciers are common.*

Sediments that accumulate in summer are usually light in color, whereas winter sediments are dark. Such layers are usually less than one-tenth of an inch in thickness. Since they typically accumulate in still or quiet water like ponds or lakes (and are thus said to be "lacustrine" in origin), Larry surmised that just such a body of water might once have existed where the road crew was now working.

---

*As he got to know more about the soils at the Gray site, Larry later revised his original assessment and recategorized the layered clays as "rhythmites," which are similar to "varves" but have no relation to glacial movements. Instead, they are sediments deposited "rhythmically" by the seasons or by major storm events.

In East Tennessee, soft black sediments are usually more recent in origin than the underlying bedrock and may date to the Pleistocene Epoch, the closest geologic age to the present one. Spanning a period from approximately 1.8 million years ago to about 10,000 years ago, the Pleistocene Epoch represents the time when glaciation covered much of the planet. It is more commonly known as the "Ice Age."

Larry and I were intrigued by the Ice Age possibility since geologic and fossil remnants from that period are very rare in East Tennessee. Erosion and climate changes have removed most of them, but when they are found, it is usually in old sinkholes, natural ponds, or caves. Larry had seen similar deposits of layered material in the western United States, where he had once worked as a mineral exploration geologist. Those western clays were definitely of Ice Age origin, and if the clays found at Gray were of similar age, then they certainly merited a further look. We considered the possibility that an old farm pond, long since filled in, might offer a more mundane explanation, but the extent of the layered clays seemed to rule against that idea.

As we pondered the more esoteric scientific questions posed by the odd occurrence of the clays, we also debated the practical questions of what should be done to stabilize the subgrade at the construction site. Sharing our ideas with Lanny Eggers, Freddie Holly, and Larry Hathaway, the TDOT project inspector, we concluded that the contractor needed to undercut the soft clay approximately five feet below the proposed roadbed grade, then backfill up to the roadway level with "shot rock"—chunks of clean, quarried limestone. This procedure had usually proven successful in similar situations.

I also decided that Larry Bolt should return to the site, and he was anxious to go back there to see what he might find in the clays. On this visit he was to observe the undercutting procedure, not only to make sure that the soft material was properly removed but also to investigate the sediments and note the scientific origin of the unusual layered clays.

I myself had once examined such clays in West Tennessee while working on my master's thesis at the University of Tennessee. Those clays were found in a pit being mined for ceramic purposes near the town of Puryer. As I parted the layers of clay into thin slabs to study them, I saw the leaf imprints of maple, gum, and oak trees. Perhaps, I thought, the clays at Gray would yield similar imprint fossils.

I had no idea at the time that they would yield something of far greater scientific significance.

2

The Discovery

The excitement and wonder of encountering new landscapes, adventures, and places of natural beauty has infected me since my childhood, when my father would take me on discovery walks in our neighborhood. I could hardly wait until the next walk, when we might find something new, like the broken eggshell of a robin or an unusual tree leaf. The possibility of a fresh discovery excited me on trip after trip.

In my adult life, exploring the wild landscapes in the American West has likewise thrilled me. Places such as Glacier National Park in Montana, the Grand Tetons and Yellowstone National Park in Wyoming, the natural arches and canyons in Utah, and the Grand Canyon in Arizona have continued to fill me with the wonders of discovery.

The same feeling would come back to me as events at the Gray road construction project unfolded. In fact, I found that those experiences would affect my associates in the same way. The power of discovery can be overwhelming and addictive.

At Gray the discoveries began shortly after the decision was made to undercut the soft clay material and allow the Summers Taylor work crew to resume its excavation of the roadbed area. Each pass of the contractor's earth-moving machinery would carry off tons of the soft gray and black clay. Most of the clay was to be taken to a waste site where the unwanted material would be piled up. Usually, waste sites are selected and agreed on by the contractor and an adjacent property owner, who might want a ravine or low area filled in. Also, waste sites are often used by the private sector for future development or to create a flat pasture area. The project contractor has the responsibility of not only grading the earth materials to make the new roadbed but also disposing of any unusable soil.

Larry Bolt returned to the Gray site on May 31, accompanied by several geologists from the Knoxville office of the Division of Geology, Tennessee Department of Environment and Conservation (TDEC). Larry had invited anyone from that office who might be interested in seeing the excavation to join him, and those who accepted were Peter Lemiszki, Martin Kohl, and Bob Price. The trip up Interstate 81 took an hour and a half, but the time passed quickly as the four geologists exchanged ideas and conjectures about the odd occurrence of the layered clays.

Looking around the site upon their arrival, the geologists were at first puzzled by the exceptional thickness and extent of the black and gray clays. In most instances in East Tennessee, clay beds of this sort do not exceed two or three feet in thickness; typically, too, they are no more than ten to twenty feet across. The clays uncovered at Gray, however, were more than fifteen feet thick and one hundred yards across. And this was just what the road crew had exposed so far. Further excavation might reveal the clays to be even more extensive in depth and area.

The geologists also noticed that some of the clay beds had a "dipping" structure. This meant that the layers were not flat but tended to curve downward into the earth, a deformation that may have resulted from a sinkhole subsidence, a cave collapse, or earthquake activity. Also, the gravel deposits found in portions of the clay bed were generally rounded, not angular in shape, which indicated weathering of some sort. Composed of "chert," a variety of quartz that is quite common in the rocks in East Tennessee, the gravel also seemed to overlay the gray clay deposit and appeared to be more recent in origin than the underlying clays. This arrangement, plus the rounded edges of the gravel, suggested to the geologists that a stream of some sort may once have flowed through this area.

The four geologists from Knoxville looked on as the construction crew removed the clay with a backhoe. Before long, they noticed some small, unusual fragments of dark brown material mixed in with the excavated clay. On closer examination, this material appeared to be pieces of bone. The men began looking at each other as if to say, "What's going on here?"

The excitement of discovery grew as the geologists tried to find additional bones. Martin Kohl and Larry Bolt yelled to each other almost simultaneously: "Come here and look at this!" In separate piles of excavated clay

The first discoveries of fossilized bones at the Gray Site were made by Larry Bolt of the Tennessee Department of Transportation and Martin Kohl of the Tennessee Department of Environment and Conservation. Shown here, the fossils included a broken jawbone and assorted other bones from a tapir. Photo by Larry Bolt.

located some fifteen to twenty feet apart, each had found a segment of bone with teeth. When pieced together, these fragments formed the perfect jawbone of some kind of animal. "I am amazed that we found two parts of the same bone," Larry later told me. "What are the odds of that?"

Their companion Bob Price added, "When we found fossilized jaw bones, I was stunned."

"Everybody was suddenly picking up bones," Larry recalled, and the geologists were soon engaged in a lively discussion about the possible origins of the specimens. Had they found the beginning of something big, or was this just an isolated deposit of bones from a modern mammal—bones that might be only a few hundred years old at most?

Returning to the office late that afternoon, Larry showed me what they had found: the jawbone pieces, plus segments of what appeared to be leg bones about six to eight inches in length and several other irregularly shaped fragments. Taking the bones outside for better lighting, we photographed them with our office's digital camera. The jawbone might have come from a dog, we thought, but it, as well as the other fragments, appeared to be very old. In particular, the heaviness of the bone pieces seemed to be evidence of mineralization.

The late afternoon sunlight gave the bones a slight yellowish cast. That color had an almost mystical aura that captured our imaginations and would grow in our minds over the next several weeks. We were photographing what would prove to be one of the most fantastic finds in Tennessee geology—except that none of us knew it at the time.

We entered the photographs on our computer for closer examination. We also put the bones under our binocular microscope, which revealed a porous structure similar to that of a sponge. We were now convinced that these bones were very old—thousands of years old, in fact—but just how old we did not know. There was not enough evidence yet to make a judgment. In addition, we had no clear idea about what kind of animal the bones came from. Some of them might even be human. It was important, we felt, to pinpoint their exact origin and significance.

The next day Larry told me about his wife's reaction to the bone discovery. "That sounds like something you would find," she had said. "You realize that no one is going to believe you, don't you?"

A few days later, Martin Kohl, another of the geologists who had visited the site, volunteered to show some of the bone fragments and teeth to Paul Parmalee, emeritus director of the University of Tennessee's McClung Museum. Dr. Parmalee's distinguished career in paleontology and archaeology included many years of research and examination of vertebrate animals from the Ice Age. In addition, he had published his findings in numerous scholarly publications. If anyone could help us identify the bone material, he could.

Bearded, silver-haired, and erudite, Dr. Parmalee showed great interest in the find and wanted to know more about the location of the site. He thought that the bones were most likely those of a tapir, a mammal that had disappeared from North America at the close of the last Ice Age some 10,000 years ago. Dr. Parmalee had identified tapir remains in East Tennessee caves, and he wanted to learn more about the soil that had surrounded the bones.

Dr. Parmalee ventured the tentative opinion that the bones were probably remains from the close of the last Ice Age event, the Pleistocene Epoch.

Top: Dr. Paul Parmalee examines a tapir bone in early June 2000. TDOT engineer Jamie Carden looks on.

Bottom: Close examination of the dark clay disclosed fossil bone fragments and teeth. Photo by Larry Bolt.

He based his conclusion on the apparent freshness of the clay and bones, as well as on earlier discoveries of mastodon and mammoth bones in nearby localities, both in upper East Tennessee and in southwestern Virginia, particularly at the town of Saltville.

"Tapir remains," he further explained, "are often associated with animal assemblages from cave and rock shelter deposits of the mid- to late-Pleistocene period in Tennessee."

He suggested that we needed to collect as many of the bones as we could to aid in their identification before the road-grading equipment moved and destroyed them. To help us, Dr. Parmalee contacted a colleague, Walter Klippel in the University of Tennessee Department of Anthropology, and an anthropology student named Marta Adams. Marta had moved to East Tennessee from the South American country of Colombia a number of years earlier. A friend of Larry Bolt's, she had initially become aware of the bone find at Gray through conversations with Larry. At the time, she was completing her undergraduate degree at the University of Tennessee and was intrigued by what Larry had told her about the discovery.

Her background in anthropology would make Marta invaluable in helping to determine whether any human artifacts were present at the site. Marta was also knowledgeable about Pleistocene vertebrate mammals and wanted to see the deposit firsthand.

The weeks that lay ahead would result in a successful salvaging effort that was unlike anything that TDOT had experienced to date. Personnel from TDOT, the University of Tennessee Departments of Anthropology and Geological Sciences, and the Tennessee Department of Environment and Conservation, Division of Geology, would supply the manpower for this busy activity. The month of June would prove to be extraordinary for everyone involved.

3

The Month of June

In early June our office was abuzz with almost constant discussion about the discoveries at the Gray road project. Larry Bolt and I bantered back and forth about what the bones might mean or represent. George Danker, another TDOT geologist, along with David Barker and Saieb Haddad, both TDOT engineers, also suggested ideas and questioned others in the office as to their thoughts about the origins of the bone deposit. Based on our own knowledge and on what Dr. Parmalee had said, the prevailing guess was this: the site had once been a pond where Ice Age animals had gathered for some reason, probably for food and water, and had drowned after being trapped in the muddy sediment.

There were other opinions as well. Some of us theorized that the site might simply have been the location of an old pioneer farm where animals had been slaughtered for food. Or perhaps it had been a small lake, formed thousands of years ago when a river meander was cut off from the main stream. Whatever the explanation, we knew that it was time for us to revisit the site and make sure that the findings were not just a one-time, isolated occurrence but actually represented a widespread deposit of unusual fossil bones.

On June 8, four of us from the TDOT Geotechnical Engineering office staff in Knoxville—Larry Bolt, David Barker, George Danker, and I—headed for Gray. It was an exciting drive, as we were all anxious to see if there were additional bones. We anticipated finding, or at least hoped to find, large bones and fossil animal skulls. We continued to speculate on what the origins of the deposit might be and wondered aloud about whether other nearby environments, such as sinkholes, could have contributed to it.

The day was warm with a slight breeze. Several white, puffy clouds skipped across the sky as we approached the road project. It was a perfect day for exploring, and when we got there, we were out of the vehicle almost before it stopped rolling.

Walking around the site, we saw piles of black lignite and some limestone boulders in the roadbed along the proposed centerline of the S.R. 75 relocation. We started sifting through the piles and immediately began to find pieces of bones and teeth. From a scientific standpoint, it was distressing to see how the construction work had disturbed the bones from their original resting places. In some spots the grading equipment had broken and even crushed the bones. If no one salvaged the bone deposit, the road project would soon destroy it. (Of course, without the road project, the bones would never have been discovered in the first place.)

We were lucky on this particular day. The road crew was at work about a half-mile away from us, thus allowing us to freely explore the excavated clay beds.

My colleagues and I busily picked through the clay piles with our fingers and pocketknives. George, a strapping blond-haired man who stands over six feet four inches tall, had a special knack for finding peculiar pieces of rock, and he always carried a hand-lens with him. "I've found something," he yelled, as he used his lens to examine a finger-sized bone fragment. Larry, meanwhile, made a similar discovery as he picked away at a large clump of blackish-gray clay. For my part, I found a mass of clay with several bones and teeth still imbedded in it. I kept this clump intact, thinking that I would take it apart and analyze it later at the office. It was obvious to all of us that we would have no problem confirming the earlier discovery of bone material.

However, we had no idea at the time about what kinds of animals these bones came from or even what body parts they represented. Most of the fragments were about two to three inches in length with diameters that were pencil-size or slightly larger. The thumbnail-sized teeth were obvious, but the animals they once belonged to were not.

Our excitement began to escalate as we continued to find well-preserved bone fragments and occasional teeth. We became increasingly sure that whatever the animals were, they were not from an old farm. The bones were noticeably heavy, indicating possible fossilization by mineral replacement,

The author found this clump of bone-laden black clay at the road project on June 8, 2000.

perhaps with quartz or calcite. Mineral replacement of bone usually takes thousands of years because of the chemical exchange required for the process.

Before long everyone had found bones and teeth. The sun was bearing down on us from the bright blue June sky, but it scarcely curbed our enthusiasm. We carefully turned each clump of clay piled up along the graded roadbed. The dark brown fragments of bone stood out readily against the dark gray and black color of the clay.

We decided to break for lunch and go to the local restaurant, the aptly named Sit-N-Bull, located in the small downtown of Gray. We looked forward to sipping ice tea and refreshing ourselves in the air-conditioning. We entered the restaurant among a lunchtime crowd that ranged from businessmen in ties to others more casually dressed, including the local service and construction people in blue jeans. A careful observer could tell that we were new to the eatery: we had to ask where to sit and what was being served. Everyone else seemed to know each other, and they clearly knew the routine and the menu by heart.

We could smell the home-cooked vegetables and cornbread the other patrons were already eating. The menu included some of my favorite dishes, such as country-style steak and gravy and fried chicken with mashed potatoes and pinto beans. Hamburgers with fries were also on the menu, as were other sandwiches such as ham and cheese and open-faced roast beef and gravy. After our morning in the hot sun, we were ready for some home-style cooking.

As we contemplated the lunch plates being served, we also talked about the dramatic things we were finding in the clay. Because of the weight of the bones, their aged appearance, and their presence in soft black sediment, we came to a striking, if still tentative, consensus: we had indeed come across a significant deposit of fossils that were probably from the Pleistocene Epoch. This meant that they might have originated up to 2 million years ago.

The surrounding limestone strata of the area dated back hundreds of millions of years. The sediments at the site were obviously not that old, but the presence of lignite indicated a far more ancient soil than any geologist might reasonably have expected to find there.

After our lunch break, we returned to the site to finish examining the clay material before heading back to the office. The ninety-degree heat was beginning to tire our little troupe, but we had a mission to find answers to the questions posed by the bones before the road grading continued.

It was important to establish that there were additional animal bones at the site and that the remains were still resting in place as they had been since the animals died. If these remains were fully articulated, with the bones intact and in place, then it would mean that the animals probably died suddenly and were not disturbed after death. This would provide scientists with an excellent snapshot into the geologic past. The fossil deposit could tell them what the environment may have been like and what may have caused the demise of the animals. For us, it was a remarkable departure from our usual routines. We were accustomed to analyzing soil and rocks for engineering properties and design purposes; looking for fossil bones was a wonderful change.

Our curiosity led us to inspect the excavation along the Fulkerson Road area, which lay just one hundred feet from our initial examination point. At this spot, gray and black clays were being uncovered, and it was easy to see their thinly laminated character. These were like the clays that

Larry Bolt had examined a few weeks earlier and that led him to conclude that they were "lacustrine" in origin.

We decided to do some exploratory work at this part of the site, and for help we called on Larry Hathaway, the TDOT road project inspector who was there that day. Congenial and cooperative, Larry had worked with me on the same road project in Unicoi County that had introduced me to Lanny Eggers. After five years of hard work on that stressful mountain road project, he and I had become good friends. I trusted Larry to give me accurate information, and he never failed to do so.

Larry was able to enlist the contractor's assistance in digging some "test pits" at the Fulkerson Road location. These would allow us to see whether any bones could be found there. We instructed the operator of the "trackhoe" (essentially a backhoe that moves on tracks like a bulldozer) to dig several test trenches about four to six feet deep. We expected these cuts to provide some quick information.

The large yellow trackhoe eased onto the graded area and twisted and turned until it was positioned in what we thought was the best location.

On June 8, 2000, the construction company dug a test pit that aided TDOT personnel in determining the extent of the soft clay deposits and their fossil content.

This put our exploratory digging operation in the middle of the Fulkerson Road excavation just before it intersected with the new S.R. 75 excavation.

Slowly and carefully, the operator maneuvered the teeth of the trackhoe's bucket into the gray clay. It cut through the soil easily and pulled up a chunk of earth that could fill two bathtubs. In one ten-second swipe, the machine was able to remove what it would have taken four men an entire day to dig out by hand. While the power of the trackhoe was amazing, it was also unnerving to think that such machines could easily have destroyed the entire fossil site in a matter of days. It was a stroke of luck that a group of geologists, particularly Larry Bolt, were at the site during those critical first few days when the soft clay was being undercut.

The trackhoe deposited its scoop of clay in a mound beside the freshly dug hole. We quickly began to look through the black clay, searching for bones and teeth. The clay broke into clumps, which were easy to separate along the bedding planes, or thin layers of the sediment, and there we were able to find fossil plant debris.

We did not uncover much in that first exploratory effort. Only some lignite and a few bone fragments and teeth turned up. We thought that there might not be much fossil material on the Fulkerson Road side, so we quit for the day. Before we left the site, I strolled around to the different exposures and took photographs. Using the office digital camera as well as my old 35mm Nikon, I concentrated on a hill of yellowish-brown, sandy clay that contained gravel, as well as the pile of clay and boulders we had examined that morning. I decided that I would take photographs each time we went to the site in order to capture the clay-deposit excavations on film in case they were later removed. At this point, it seemed quite possible that nothing else would be found at the site and that our excitement would fade away. However, it was also possible that we were on the brink of major discoveries, and I was determined to make a visual record of these events as they unfolded.

On June 14, Larry Bolt returned to the site to study the effects of undercutting the clay. For this trip, Larry enlisted the assistance of Marta Adams, who could help determine whether the bones were human or nonhuman. This step was vital because TDOT is required by state statute to make such a determination whenever bone material is uncovered in a road-project excavation. If the bones are found to be human, they would have to be professionally studied on-site and then removed for protection and continued research. By law, TDOT must halt any road construction

Anthropology student Marta Adams helped to expose and salvage portions of a tapir skull on June 14, 2000.

in the immediate area of a human bone find until the bones are removed. In most instances, the bones prove to be from Native American burials, but sometimes they turn out to be from more recent disposals.

Upon arriving at the site, Larry and Marta found the construction crew proceeding with the excavation of the clay down to the proposed roadway grade elevation. Larry saw that lignite and numerous bones were being exposed as the grading equipment cut through the clay along the mainline of the project at the Fulkerson Road intersection. Larry and Marta quickly set out to salvage whatever they could between the passing rounds of the contractor's machines.

As an earth-moving pan or a bulldozer made a pass through the site, Larry and Marta quickly followed behind it, identifying possible bone fragments and removing them. They were astonished to see more and more bones surfacing with each successive pass of the equipment. The rust-colored bones stood out in sharp contrast against the clay. Before the next

Fossil-Hunting Tools

Paleontologists—the scientists who study life forms from past geological periods by examining fossil remains—use a variety of tools in their field work. Although much of this equipment is fairly commonplace, a few instruments are not exactly household items.

A rock hammer and some type of chisel are the most basic aids to paleontologists as they split layered rock, such as shale and shaley limestone, in their search for fossils. Often a combination tool called a chisel hammer or mason's hammer is used. Its head, similar to that of a claw hammer, consists of a blunt hammer on one side and, opposite that, a flat chisel. Another commonly used combination tool is a pick hammer, whose sharp point, in experienced hands, can split rock with great precision. A pry bar may also be useful for uncovering large, rock-embedded specimens.

For excavations in softer, clay-like strata, such as that found at the Gray Fossil Site, small shovels of the sort used by the military and by some gardeners are handy. In addition, instruments with flat, wide blades—pocket knives, putty knives, and trowels, for example—can be used to dig and peel away the clay from the fossil. In situations involving delicate fossils or those with minuscule details, dental tools with long, thin pick ends may help remove the encrusted soil and rock. Even an ice pick will do the job. Tweezers are also handy for picking up very small, delicate pieces of fossils.

Still another tool that is often quite useful is a sieve, which allows one to sift through fine soil or sand. A sieve typically consists of a piece of wire mesh or screen fastened to a wooden or metal frame, which can be either round or square. Sieves are available in different sizes for sifting particles of varying coarseness.

Fossil hunters frequently employ various small brushes to clean away rock or clay material. Called "dust brushes," they usually include an ordinary tooth brush, a quarter-inch-wide paint brush, and a wider, two- to three-inch paint brush. Another useful tool is a plastic squirt bottle that can produce a light spray or small stream of water for cleaning fossil surfaces in the field. Detailed cleaning and preparation processes, however, are usually best completed in a laboratory setting.

A magnifying hand lens is a must when looking for fossils. These can vary in magnification from 3x to 10x and even 20x. A good lens can reveal

sweep, they would gather as many fragments as possible. "It became extremely exciting when the machine operators were looking out for bones, and as soon as they would spot some they would stop and call us," Marta recalled.

Larry dug up the pieces with his old army shovel while Marta used a trowel. But no matter how quickly they tried to work, their progress was tedious and slow. They unearthed the leg bones, about twelve to fourteen tiny details such as the structure and form of shells and bones (what scientists call their morphology) as well as the mineralization structure of fossils. Also, a binocular microscope, often standard equipment in a field lab, can be invaluable in examining minute details of fossil material.

Additional field equipment for fossil-hunting expeditions may include a compass, a good topographic map, some length of measuring tape (preferably metric), cloth bags or plastic zip-lock bags, old newspapers for wrapping specimens, a backpack or knapsack, and, of course, a camera.

In recent years, advances in the technology of geophysical research have made it possible for scientists from universities and other institutions to do remote, non-invasive searches for fossils in clays and sands. One technique involves resistivity measurements and is accomplished by running a direct electrical current through the subsurface of an area and measuring the resulting changes in voltage as the current meets resistance within the underlying layers of soil and rock. Another technique, measuring seismic reflection, involves sending seismic waves (produced by a large pounding hammer or small explosive blast) from the surface into the underlying strata, which results in some of the waves being reflected back to the surface as they encounter geological discontinuities, such as bedding planes, faults, cracks, or even a layer of fossil bones. Receptors record these waves and their travel time and can tell a scientist much about the geological makeup of a given subsurface. Still another technique uses ground-penetrating, high-frequency electromagnetic waves, or radar, to probe subsurface conditions and produce cross-section images of the geologic strata and any anomalies that may be present, including fossil bones and skeletons.

It is important to remember that fossils are part of the earth's history and provide important facts and clues about life forms from the distant past. They are a finite resource, and the random digging and collecting of fossils just for the sake of collection are not encouraged. By reading resource materials such as guidebooks and textbooks, one can better understand what to look for and what the fossil may mean in the larger context of geologic history. Knowing what fossils represent greatly enhances one's field experience. ■

inches long, of a probable tapir. Exposed to the air, the color of the bones changed within minutes to a darker shade of brown.

Larry then found what appeared to be a skull the size of a small melon, about ten to twelve inches long. It was covered in dark clay, and later, after Larry brought it to the office, it required numerous swashes with a brush and water to get it clean enough to determine what it was. It proved to be the skull of a relatively small tapir, perhaps a young adult.

In the face of these important discoveries, Larry and Marta became increasingly frustrated as they saw numerous articulated bones in the clay being damaged with each pass of the scraper equipment. It was hard for them to watch as the bones were destroyed or broken loose from their clay tombs. The pair tried to keep up with the excavating machines, but they could not remove the bones fast enough. There were just too many of them. Larry took some photographs of the bones before removing them. In the midst of their searches, Marta told Larry, "It's a dream site for any researcher." Later she recalled, "I immediately knew it was a special find, and we could not let the bulldozers destroy it."

Larry and Marta quickly amassed a remarkable array of bone debris amid the passing earthmovers and bulldozers, enough to fill two cardboard boxes normally used for storing files. There were pieces of skulls, numerous leg and toe bones, vertebrae, and ribs. It seemed that everywhere they had looked during the excavation process, bones had turned up. The discovery at the Gray site now appeared to be no ordinary Ice Age deposit, and Marta was confident that the bones came from animals, not humans.

Larry returned to the office that day carrying one of the cardboard boxes filled with bones, and he brimmed with excitement as he announced what he and Marta had found. Marta, meanwhile, had taken the other box of bones to the university's anthropology lab for cleaning and identification.

Larry also expressed concern that the contractor's road-grading efforts were destroying the bones. He and I had several discussions, some even argumentative, about what options to pursue in salvaging the bones without delaying the contractor's operations. We considered going to the site on weekends, when no construction work was being done.

We also discussed the idea of asking the project engineer, Lanny Eggers, for help. There was no precedent for this situation. Highway construction in Tennessee had never before uncovered this many fossil remains of vertebrate animals. Crews might encounter an occasional mastodon tooth or bone in an old sinkhole or farm pond, but there was nothing to compare to this deposit. We were entering into uncharted territory.

For the next several weeks, we continued to visit the site to look for bones and devise a plan for their recovery. The project inspector, Larry Hathaway, became more and more interested in what we were finding and began looking for specimens himself. Once he saw what the fossil material looked like, he became one of the best "bone hunters" on the project.

On June 22, 2000, the author found these water-filled "potholes"—evidence that local residents were digging in the area after working hours.

As we continued to weigh the contractor's needs against the interests of science and preservation, another problem developed during the middle of June. Local residents, having observed our digging and salvage activity, assumed that we were finding and removing Native American bones and pottery. As a result, a number of them started making after-hours searches for artifacts. Numerous "potholes" began showing up daily in the grade of the roadway project as evidence of where local diggers had been the previous evening. One day, I found that the site had been so scoured the night before by amateur bone hunters that the ground looked like a battlefield after a barrage of mortar shells. Craters three and four feet in diameter were everywhere.

We knew that between the destructive sweeps of the grading equipment and the takings of diggers, many of the fossil remains would be lost to researchers. And with them would vanish important evidence of the site's geologic history. Our concerns that something needed to be done to protect the bones were heightened by our growing perception that the site might well be a true scientific treasure.

We could now see that this area was unusually rich in clues to an exotic prehistoric landscape. The bones we had already found were providing invaluable information about the context of the ancient environment: the condition of the water that once was there, the sediment in what had been a pond bottom, the plants that had surrounded the pond, the types of animals that had lived there, the ways in which the animals died, how the sediment was deposited into the pond, and much more. It was all there, preserved in one nice package—except that a highway was being built right through the middle of it.

Word spread throughout the local community about the bones, and the nighttime digging escalated at the site. The workers were still four to six feet above grade elevation in their excavation, so, outside of any obvious safety concern, the contractor did not seem worried about souvenir hunters coming to the project after each day's work was completed. A few extra holes in the ground would not hinder the contractor's operations at this point.

Vanessa Bateman, a TDOT geologist from the Nashville office, and I visited the site on June 20. As a geologist, she was especially intrigued by word of the fossil discoveries, so I had invited her to join me on a bone-hunting expedition.

Larry Hathaway met us there and described where the bones had been showing up during the grading operations. We proceeded to the centerline area of the new roadbed where the gray and black clays were exposed.

We began to look around. Even on the surface we found numerous pieces of bone. Larry told us that on the day before, he had uncovered a number of bones by randomly piercing the ground with a shovel. He had no idea as to what animal they came from, but he thought that we should see them. He showed us a small box full of specimens, most of which measured about five or six inches in length. They included rib pieces, vertebrae, and a few teeth.

I decided that we should repeat Larry's experiment with the shovel. I was skeptical that we would find anything, since the process seemed so random, but Larry was confident that something would turn up. Then, on our first try, Vanessa, Larry, and I met resistance about ten inches below the surface. What had we hit with our shovels? I started to sweat, as much from anticipation and excitement as from the heat.

We began to dig slowly and carefully to avoid damaging what might be a significant fossil. We soon uncovered part of a vertebra, which we brushed off while it still lay buried in the clay. We moved laterally along the path of what appeared to be a backbone until we had uncovered a good portion of it, careful to leave it in place. We began to expose rib bones along the path of the vertebrae. The bone was a rusty tan color at first, but within an hour or so it changed to a dark gray. We assumed that this was caused by oxidation that occurred when the fossil was exposed to the air.

Vanessa used the shovel to probe the soil once more. And again, on the first try, she hit bone. Together we began the slow and careful excavation process, removing the clay surrounding her find. After about thirty minutes, we had uncovered about ten vertebrae in a row, all articulated. We continued working and reached what appeared to be the end of the vertebrae—perhaps a tail. We had little doubt that there were more skeletal remains within that little patch of ground. It seemed logical that the animal had died in this spot and that the rest of its bones would be there.

Shortly after uncovering the backbone, we calculated where the head or skull portion should be and began digging there. Slowly I eased the shovel blade into the clay. Again, after about six or seven inches, I hit something hard. I gently withdrew the shovel and began to use a putty knife to remove the overburden soil. The bone material quickly came into view.

Sure enough, at the very last vertebra, exactly where the backbone would attach to the skull, was a larger piece of bone. The material seemed to be fractured, but it still held together. I carefully uncovered the bone until I had reached a tooth. Then I knew I had a skull or, at the very least, a jawbone.

It was soon apparent that it was an entire skull, with the lower jawbone still intact. We uncovered the skull from the back portion to the front teeth. At this point, only the left side of the skull was revealed, showing in a bas-relief form. The skull was approximately fourteen inches from front to back and about eight inches from the top to the bottom of the jawbone. The skull mass resembled a small melon, like the one that Larry Bolt had found a week earlier. Again the bone was a rusty tan color that stood out distinctly against the clay.

Vanessa cleaned the backbone while I uncovered the skull. We wondered about the sort of animal these fossils represented. I recalled that Paul Parmalee had identified our earlier discoveries at the site as the remains of tapirs. This was probably the same kind of animal, we thought to ourselves, as we continued our excavation.

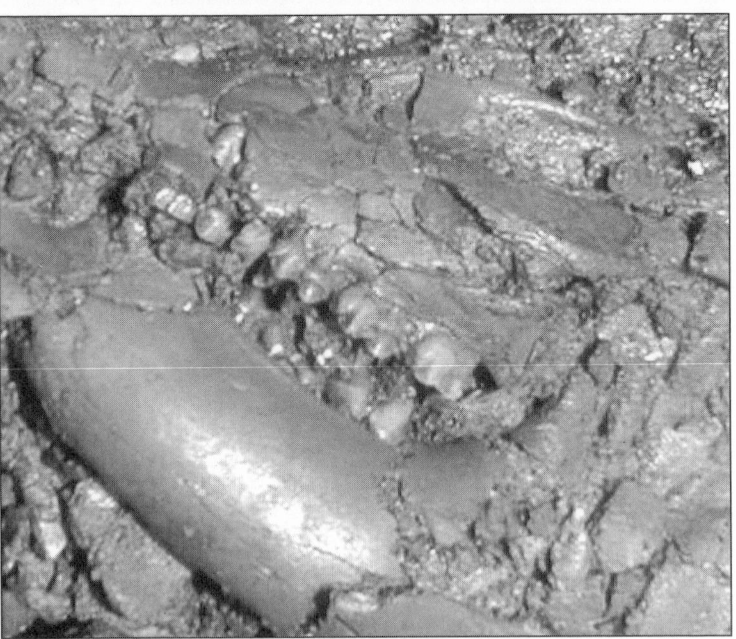

Top: The author found this tapir skull on June 20, 2000.

Above: The teeth of the tapirs found at the construction site were in excellent condition.

The teeth were now showing up as we cleaned the skull with water that had collected in a nearby hole from the previous night's rain. The teeth were remarkably well preserved. Except for a grayish cast, the enamel was perfect. Picking away at the clay, we soon found the front leg bones beneath the skull. Although we did not expose the toes that day, we were confident that the entire skeleton was there.

By mid-afternoon we had uncovered what we guessed to be almost the entire tapir. The bones were indeed similar to those previously identified by Dr. Parmalee.

The sun was hot, and we were becoming dehydrated. We knew that we had to remove our discovery before the "night shift"—the local souvenir hunters—arrived. They were sure to take anything that was left exposed. Thinking that the bones were Native American artifacts, the locals had no idea that these might in fact be relics from the Ice Age. We obviously were not certain about their origins ourselves, but the signs were pointing in that direction.

I decided to make some photographs to document our find. We gently brushed water over the bone surfaces so that they would stand out in the pictures. A wet surface tends to bring out details, and these bones had details galore. Using both a digital camera and my trusty Nikon, I took wide shots of the skeleton as well as close-ups of the skull, backbone, and teeth. I included overhead views and side views. After the photography session, Vanessa and I carefully boxed up the remains to take back to Knoxville.

We were now utterly convinced that a large deposit of animal remains was buried here, and that this would prove to be a fossil site of considerable scientific importance. It seemed that we could dig in almost any spot and find a bone. We had come upon an ancient animal graveyard that demanded respect.

But what could we do to protect the site? We would have to convince our TDOT administrators that the road excavation should be stopped. We believed that if only we could find the remains of something big or dramatic, such as a prehistoric elephant, then this would gain enough attention to halt the roadwork until we could make a comprehensive geologic assessment of the area.

My dilemma was this: should I alert the TDOT management now or wait until something big turned up? On June 21, I called my supervisor, Len Oliver, manager of the Geotechnical Engineering Section in Nashville. I told him about our findings thus far and about the potential dangers to the fossil deposit of continued roadwork. Advising against an immediate work stoppage, however, I suggested that we continue to examine the clay sediment in hopes of finding some larger animal remains. Len concurred.

On June 24, Vanessa and Marta returned to the site to resume the search for bone material. It was a Saturday, and the contractor's earth-moving machines were idle. Vanessa and Marta intended to remove any bones they could find before grading resumed. They randomly picked a spot to dig, just as Vanessa and I had done previously. Immediately their shovels hit bones, which they began to uncover. In the construction crew's absence, Vanessa and Marta took their time since there was no danger from heavy equipment lumbering about.

Using small trowels and putty knives, they spent most of the day uncovering the skull, backbone, and ribs of another animal. The skull had most of its teeth, and the bones were articulated. Vanessa and Marta carefully kept the bone pieces together in the clay matrix, preserving some of the surrounding material for context. Again, it was a hot and humid day, and they both became dehydrated during the painstaking removal process.

The specimens looked like tapir bones, but the flatness of the skull and the seemingly extensive number of vertebrae, as compared to known tapir skeletons, gave them a few doubts. They speculated that it might be the remains of a horse of some kind. However, closer examination at the university's anthropology lab a few days later revealed that the animal bones were indeed those of a tapir.

When Vanessa and Marta called me about their latest find, they expressed concern that something needed to be done quickly to stop the road excavation. They were convinced that more bones lay beneath the surface. I told them about my conversation with Len Oliver and our agreement that we should continue looking for something even more dramatic than the tapir bones.

Among the state's scientists, enthusiasm was building over the Gray site. Geologists with the Tennessee Division of Geology and the Tennessee Department of Transportation, along with members of the University of Tennessee Department of Geological Sciences, began to frequent the site several times a week. It was becoming increasingly clear to them that this site held secrets to the past of East Tennessee that had never been revealed before.

One of the scientists who joined the investigations was Mike Clark, a professor of geology at the University of Tennessee. He was a specialist in the study of landforms and the processes involved in the formation of the earth's surface, a branch of science known as geomorphology. In mid-June, he had begun to analyze surface deposits at Gray. His research entailed the minute measurement of the overlying layers of soil and gravel deposits. He wanted to determine how the present surface related to the boundary that separated the underlying gray clay sediments from the overlying cherty gravel and reddish-brown soil. Such contact between two contrasting layers of sediment is referred to as an unconformity, and its occurrence at Gray was especially intriguing to Dr. Clark.

The geomorphic analysis required Dr. Clark to describe the different layers of strata that could be found at the site, beginning with the highest horizons at the top of the cut slope made by the roadway excavating equipment. For several weeks, whenever I visited the site, I would see Dr. Clark hard at work digging out a trench profile in the soil cover and using a range rod (a surveying tool for vertical measurements) and a hand level to take readings. He enlisted a volunteer student from nearby East Tennessee State University to help with his work.

With Dr. Clark, Dr. Parmalee, Dr. Klippel, and Marta Adams all engaged in field study along with the TDOT staff, the site began to resemble a research encampment out in the bush—except that we were in the middle of a residential area next to a school. The days became increasingly exciting as we encountered new fossil bones on each visit. It was difficult to keep up with what was being found and who was in possession of what. We were also trying to keep out of the way of the Summers Taylor construction crew. They were patient with all the commotion and the visitors to the construction site. At this point, the contractor was still adhering to the work schedule but was voluntarily staying out of the fossil-bone area in order to let TDOT decide what needed to be done.

On June 26, I brought Don Byerly, a geology professor from the University of Tennessee, to the site. Although I had once been his student, I actually knew him even before I entered college. When I was a teenager, he oversaw the Explorer Scout Post in Knoxville of which I was a member. We had hiked numerous trails in the Smoky Mountains together and camped in various wild areas of East Tennessee, including some caves. Because of our long friendship, I knew he would appreciate the opportunity to see this unusual fossil deposit for himself.

It was another bright, sunny day. Dr. Byerly and I arrived around midmorning and immediately found a woman and her ten-year-old son digging in the gray clay. I approached them and asked what they were looking for.

They were digging up clay with a shovel and sifting the material through their fingers like prospectors hunting for gold. They carried a white five-gallon plastic bucket for storing and carrying their finds. Standing in the middle of the excavated area, they had mud caked on their feet and hands.

The mother told me they were digging for the remains of a turtle. They had been there the night before and had found four turtle shells in the clay. This was the first face-to-face confirmation of our fears that local residents were coming onto the site at night and removing fossil material. "Do you have permission to be here?" I asked.

"Do we need it?" the mother asked back.

"Yes," I told her, "the contractor is responsible for the site, and you'll need to check with the project superintendent for permission." I described the superintendent's red pickup truck and gave her the contractor's name.

One of several turtle shells discovered at the site in June 2000. The penny next to it indicates its size.

"I understand," she said, and immediately packed up her things, along with what appeared to be a turtle shell. Her son proudly carried the fossil back to their car. I wanted to ask for their names, but I felt that I had warned them sufficiently. They did not offer to show me their find, but I could see that it was about palm-sized and dark brown. It did not look entirely whole.

After the mother and son left, Dr. Byerly and I examined a layer of sandy gray clay in the middle of a new excavation at the intersection of Fulkerson Road. It was what geologists call a "festooned fluvial deposit." Originating from a fast-flowing stream, it was a sandy sediment whose bedding was bowed or curved from the horizontal. Rounded pebbles of chert, similar to the gravel first noted by the visiting geologists on May 31, were also apparent in this exposure, as were small pieces of woody material and an occasional bone fragment.

The number of carbonized plant fragments in the sand and clay was particularly impressive. Most of the plant pieces were black and twig-like, with lengths ranging from an inch or two to more than ten inches. The fragments retained the impression of the original surface bark, and some were flattened as though they had been pressed by a heavy weight. These fragments caused us to wonder about the origins of the sandy deposit, which we tried to relate to the surrounding black clay material. Its appearance was unlike that of the adjacent layered clays, which appeared to be dipping at a moderate angle and were much darker in color. We knew that the lighter-colored sandy soil had originated in a different kind of environment from that of the black clay. Perhaps the sandy material and gravel represented the deposits of a fast-moving stream that had once flowed into a larger body of quiet standing water like a pond. This light gray sandy area also contrasted noticeably with the thinly laminated clays found on the Fulkerson Road portion of the site a hundred feet away, some of which were perfectly horizontal.

As we continued examining the clays and working our way toward the top of the deposit where the more exposed, weathered, and oxidized material was located, we began to find larger bone pieces. I was analyzing the weathered clay zone when I noticed a bit of tooth enamel protruding from the ground. Dr. Byerly, some fifty feet from me, was still inspecting the sandy material and finding pieces of black woody lignite.

I quickly dug around the enamel piece with a hand trowel to uncover it. I was not sure what it was, but it looked like part of a skull. It appeared

Freshly exposed in late June 2000, this laminated clay contained plant debris, including leaves, within the blackish layers.

to be from an animal much larger than a tapir, and despite a sudden rush of excitement, I tried not to hurry as I exposed it. What I had found was a very large tooth and a piece of a jawbone. The tooth jutted out from the bone by more than two inches, and it was more than an inch wide. It was almost four inches from the tip of the tooth crown to the end of the tooth root that stuck out from the bottom of the jawbone. The crown was a white to whitish-gray color, with a polished sheen.

I looked around to see if anyone might be watching us. No one else had found evidence of an animal of this size, and after my earlier encounter with the souvenir hunters, I was concerned that someone unassociated with the discovery might damage or disturb the area where I had just made the find. This might well be the "big" discovery we had been hoping for.

I motioned quickly for Dr. Byerly to join me. We both stared intently at the big tooth but had no idea what animal it came from. Assuming that this was indeed a Pleistocene Age deposit, our best guess was a giant ground sloth. I was sure that this fossil was a significant discovery, and we soon dubbed it simply "the big tooth."

The author found this jawbone and large tooth on June 26, 2000. It was initially thought to have come from a giant ground sloth.

While we found a few more pieces of bone nearby, they were similar to what we had seen in earlier searches. Nothing matched the big tooth. We walked over the site again, scanning odd clumps of clay and fragments of rock. Amid the black and gray clay, we noted the presence of some large limestone boulders, some as big as eight to ten feet in diameter. These boulders were new to me, as the construction crew had just uncovered them a few days earlier.

The boulders raised fresh questions about the geological history of the site. We wondered about what natural processes had placed these huge rocks within the layered clays. Perhaps, we thought, they had resulted from the weathering of exposed rock surrounding the lake or pond that might once have been there. Whatever the case, they definitely formed a new piece of the puzzle.

By 3 P.M., it was time for us to leave the site and head back to Knoxville. We were tempted to stay longer, but we had commitments in the city. As we were leaving, I decided to pick up a sample or two of the woody lignite since I wanted to show some of the people at the office what it looked like. I was near the intersection of Fulkerson Road and the new S.R. 75 centerline when I noticed what appeared to be a piece of weathered woody lignite, resembling a short section of fence post. It was yellowish and rusty in color and about three inches in diameter and seven inches long.

When I lifted it, I realized that it was more than just a piece of lignite. At five to six pounds, it was quite heavy in relation to its size, and it was unlike any of the lignite pieces we had found earlier. Noticing an unusual crosshatched pattern at both ends, it occurred to me that this might not be a plant fossil at all: such material would likely display concentric rings in cross section, not crosshatching. But I was not certain of that, either. The piece had a striped lineation that ran lengthwise down its surface in a bark-like pattern. I looked around for similar specimens, but this seemed to be the only one of its kind. Petrified wood perhaps? It was either that or some kind of animal bone.

At the office the next day, I showed Larry Bolt and George Danker what I had found during my trip with Dr. Byerly. We took the specimens into the kitchen to clean them. The jawbone and tooth cleaned easily, and everyone was impressed by the size of the tooth. Our office secretary, Nancy Chadwell, asked me questions about its origin, but I had to admit my uncertainty. I told her that I would ask Dr. Parmalee and Dr. Klippel to identify it.

The "petrified wood" proved harder to clean than the tooth and jawbone. It had a tendency to crumble along the edges. While some of the clay flaked off easily, the mass was rather splintery, and portions would break away when we tried to wash it.

Martin Kohl at the TDEC Division of Geology office came over to see the pieces at my request. He thought that the log-like piece might be a portion of some type of elephant tusk, either from a mastodon or mammoth. He had no opinions about the origins of the big tooth, but he was excited that the site was turning up such prizes. He wanted to join the efforts to look for more.

After a call from Larry, Dr. Parmalee and Dr. Klippel came over later that day. They were amazed by the tooth and jawbone and bewildered by the log-like piece. They conjectured that the tooth might have come from a giant ground sloth, but they were not sure. Enameled teeth were not characteristic of that animal. They agreed that the "petrified wood" might be part of a tusk. If so, it was very weathered and needed to be handled carefully. Additional research might reveal what it actually was.

Pieces of elephant tusk found at the project site. Note, at top, the tip of a tusk.

The ends of the fossilized elephant tusks revealed a crosshatched pattern.

Feeling more than ever that we had stumbled across a major deposit of vertebrate animal fossils, Dr. Parmalee and Dr. Klippel were anxious to find additional specimens. They offered to help clean and analyze the bones at the university's anthropology lab. We handed over some of the tapir bones, as well as the skull Vanessa and I had unearthed on June 20. We decided to keep the big tooth for a while longer to clean it and make photographs with the digital camera.

As we were in Knoxville fussing over my latest finds, we were unaware that on that same day, June 27, additional discoveries were made at the site that would help decide the fate of the road project and the gray clays of Gray, Tennessee.

Marta Adams was at the site that day with Dr. Klippel and Dr. Parmalee, excavating soil near the proposed roadway ditch line when she saw something protruding from the clay at the bottom of a freshly excavated slope. Digging in this spot, she soon uncovered a large femur-like bone, some two feet in length and several inches in diameter. "This is a large animal," Dr. Klippel told her. "This may be what you have been looking for."

Before long she hit an even larger bone, which she partially uncovered. She knew she had made a significant find, but it was getting late. She boxed up the first bone to take with her back to Knoxville, while covering up the other bone with clay.

Thinking that she may have discovered the leg bones of an elephant, Marta called Larry Bolt that night about her discoveries. She told him that the second bone should be unearthed as soon as possible. Larry in turn contacted me, and the next day, Larry and I traveled to the site with Marta and her husband, Tony Underwood, who came along to help out. The day was warm and muggy, and a light rain had developed early that morning. By the time we arrived at the site, the air was thick with humidity, and the gray clouds were becoming darker.

Locating the spot where the big bone was found, we slowly began to excavate the clay. At first we used small putty knives and then switched to shovels before we realized that these tools were inadequate. Project inspector Larry Hathaway asked one of the contractor's backhoe operators to help us. With the agility of a dentist, the operator adroitly removed layer after layer of clay around the bone. His delicate maneuvers with the backhoe amazed me.

This was no ordinary bone. It measured about three feet long and about eighteen inches wide. It was several inches thick and curved in shape.

This large bone, a portion of an elephant pelvis, was discovered by Marta Adams on June 28, 2000. Because of its considerable weight, it had to be removed in sections.

We knew that we would have a hard time lifting it once it was uncovered. The pace of the digging became quicker as the air thickened and the sky grew darker. With rain on the way, we needed to hurry.

As the backhoe cleared away the surrounding soil, we used trowels to remove the soft, wet clay that stuck directly to the bone. Several times we feared that we had broken the bone as we dug, only to find out that it remained intact beneath the clay.

Then the rain started. Big drops of water fell while we frantically dug to free the bone from the earth. The clay had sealed the remains of this beast for eons of time. Once exposed to fresh air, the light rusty brown color of the bone slowly turned to a dark gray-brown.

The rain came down harder, running down our faces and making it difficult to see. Nearby, Bob Price and Martin Kohl with the Division of Geology were also struggling in the rain, trying to uncover yet another tapir. It seemed as though a tropical storm had hit the site.

As we attempted to lift the big bone, it cracked into a number of smaller pieces, forcing us to remove it bit by bit. We could not leave the fragments at the site, as there was no way to protect them from vandals. Slowly we placed the pieces into two black plastic garbage bags, which we estimated at about 120 pounds each once we had filled them.

The rain came down harder, accompanied by lightning and thunder. The torrents poured off of us as though we were standing under a waterfall. The clay was becoming slick now. Several people at the site tried to run to their vehicles and abruptly slipped as if they had been scrambling on ice. "We could barely walk in that muddy mess," Marta later observed.

As the rain showed no signs of stopping, we decided to call it a day. Taking our time, we eased the bags containing the big bone pieces into our office's Jeep Cherokee and Marta's pickup truck and then cleaned our digging tools. Meanwhile, Marta's husband, Tony, had found what looked like a carapace, the top of a turtle shell. It was dark brown and came out of the soil in pieces. I took photographs of it, as I had also done with the big bone.

We knew now that we had found the fossils that would stop the roadwork long enough to give us time to evaluate the situation. Despite the messy weather, it had been an exciting and successful day of discovery. "We all felt such a sense of accomplishment!" Marta remembered. "We knew all the work we had done up to now had finally paid off. We had finally found the missing piece to the puzzle."

The size of the bones indicated that we had found our prehistoric elephant, and indeed this would be confirmed a few days later. What we had lifted from the ground that day would be identified as a portion of a pelvis bone belonging to a mastodon or mammoth. (We tended to think of it as a mastodon, but we were unable to confirm this because we never found any teeth, which would have given us more precise clues.)

Unbeknownst to us at the time, another major fossil find had turned up at the site on June 27, the same day that Marta had first found the elephant bones. The discoverer in this case was Rick Noseworthy, a TDOT biologist and environmental coordinator, who was there to check on erosion control measures involved in the construction project. While poking through some freshly excavated clay, he came across an odd, football-shaped black mass. From its size, he thought that it might be a large piece of fossil wood. He engaged Keven Brown, another TDOT biologist, to help him look for additional fossils. They examined numerous black clumps of

A piece of the elephant pelvis bone found on June 28, 2000.

clay and found several small bones that were later determined to have come from tapirs.

Over the following weekend, Rick scraped, needled, brushed, and pried the black clay from the odd-shaped mass he had found. After a few hours, an astonishing shape began to appear. It was somewhat flat and had a rough, bony surface with two silver dollar–sized indentations on top, which turned out to be eye sockets. Rick had found an animal skull.

Rick thought at first that it might be the deformed skull of a tapir, but its one slightly protruding tooth was different from the tapir teeth found at the site. Also, the skull narrowed toward one end, giving it an elongated triangular shape. After looking through some reference books, he soon realized that this was not a tapir but something quite different. Rick contacted us about the skull that following Monday, and we arranged to have Dr. Parmalee come to our offices to look at it.

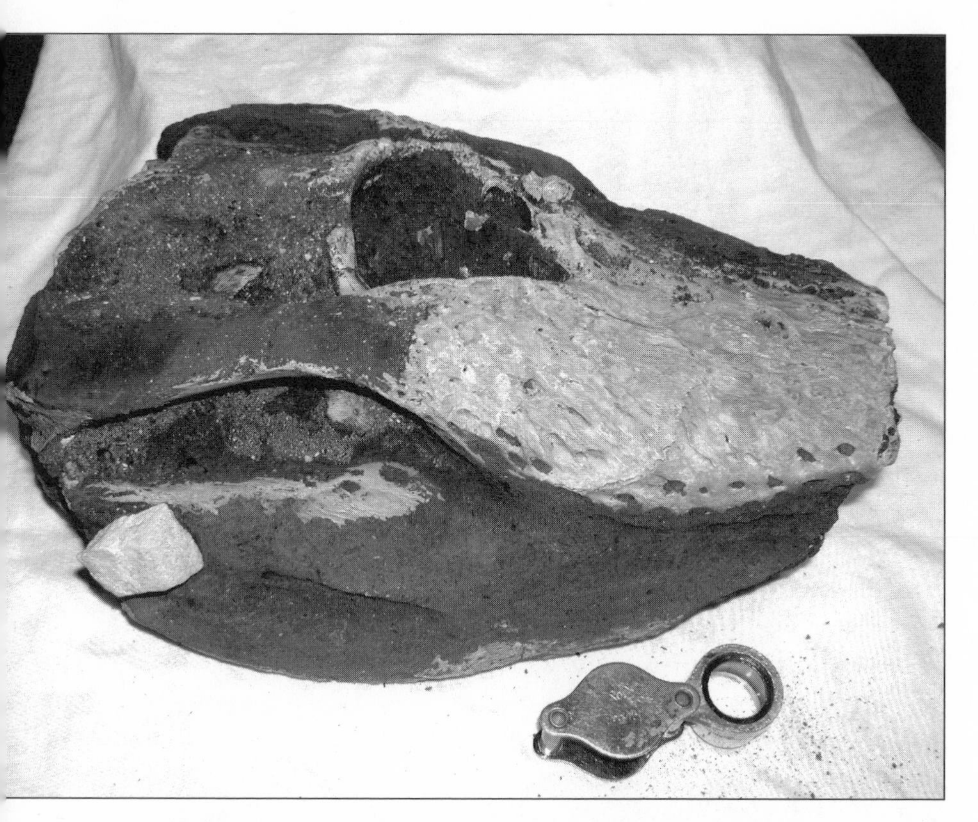

The alligator skull found at the site by Rick Noseworthy, TDOT environmental coordinator. Note the eye socket near the top of the skull.

Within an hour of our call, Dr. Parmalee arrived, accompanied by Dr. Klippel. When he saw Rick's discovery, he simply stood there motionless. A light breeze could have knocked him over. Then, very slowly, as if he could not quite believe the words coming out of his own mouth, he told us that what Rick had found was the skull of a crocodilian animal, most likely an alligator. He repeatedly asked where the fossil had been found. "At the Gray construction site," we repeatedly told him.

Dr. Parmalee believed that this was the first fossil alligator ever found in Tennessee, and its discovery began to alter his previous thinking about the age of the deposit. This meant that the bones we had found might well be much older than the 10,000 to 20,000 years we had previously thought. "The remains of alligators, aquatic turtles, and certain other taxa,"

Dr. Parmalee explained, "suggest a warmer climate when these remains were accumulating in what appears to have been a water-filled sinkhole or cavern." A warm climate, according to Dr. Parmalee, indicated something other than Ice Age origins.

At the time, Dr. Parmalee was not sure how much older the fossils might be, but he believed that checking with experts at other universities could provide some answers. The discoveries had set "the paleontological ball rolling," as Dr Parmalee put it, and its outcome was unknown at the time. Discoveries were at hand, and more were sure to come in the future.

In a glorious month of discovery, we had found tapir bones and teeth, elephant bones, the mysterious "big tooth," and now an alligator skull. The fossil deposit had become quite a complex puzzle, and it seemed that the more we found at the site, the less we knew about what we had found. But the prospect of obtaining precise answers—as well as even more fossils—was exhilarating. In a spirit of friendly camaraderie, we gave Rick Noseworthy some new nicknames: "Crocodile Noseworthy," "Rick Dundee," and the "Alligator Man."

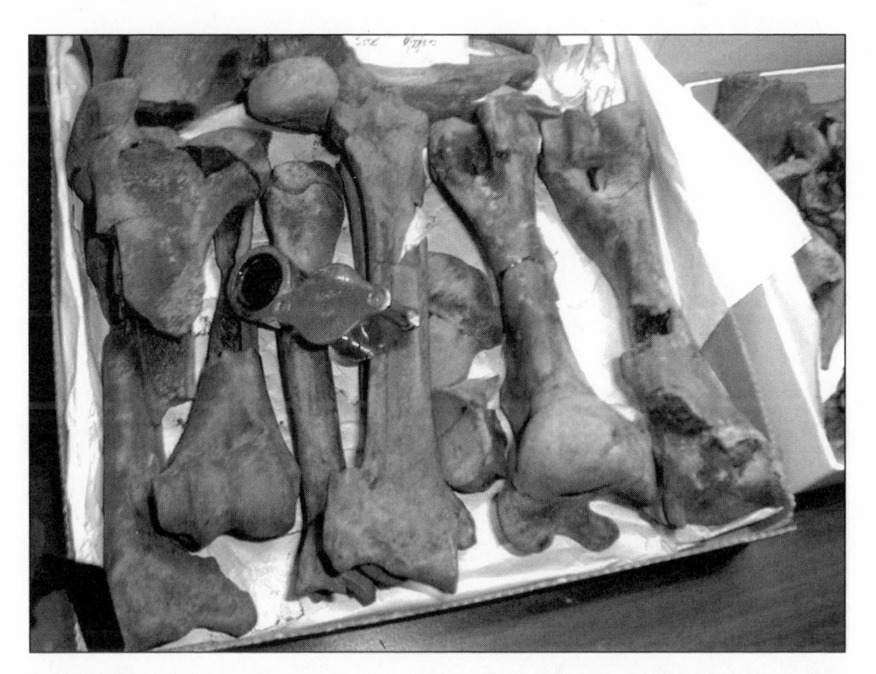

Tapir leg bones were plentiful at the site, and many of those were examined at the office of Dr. Paul Parmalee of the University of Tennessee's McClung Museum.

Dr. Parmalee compares a cast of a tapir bone, dating to the Ice Age and found in an East Tennessee cave, with the partial jaw of a tapir (in his right hand) found at the Gray road project.

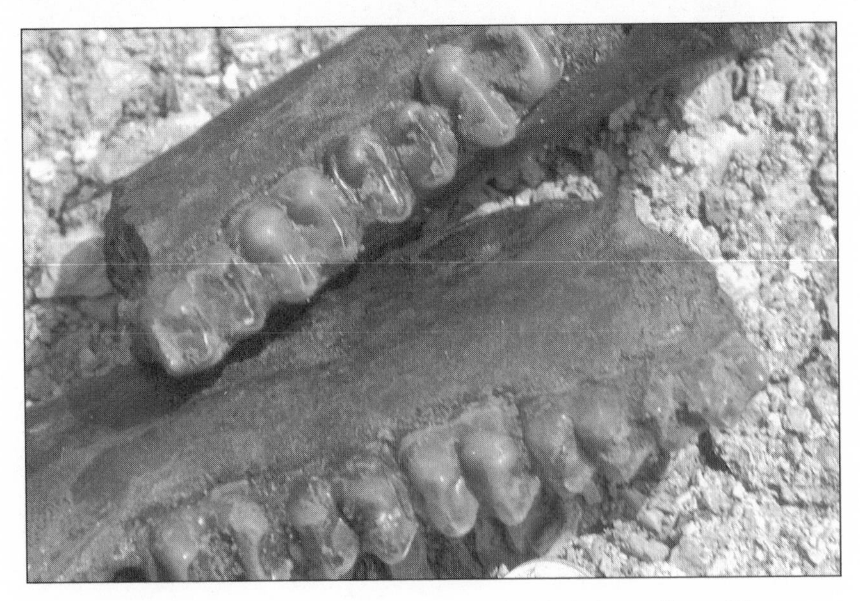

Pieces of a tapir's mandible (or lower jaw) found in the black clay portions of the roadbed at Gray.

Larry Hathaway, TDOT road inspector, became one of the best bone hunters and finders at the road construction project.

The fracturing of this tapir skull most likely resulted from the road grading work at the construction site. Photo by Vanessa Bateman.

But as we celebrated our finds, we also knew that it was only a matter of time before the news media found out about the fossil site. And once they showed up, it would be impossible to keep curiosity seekers out of the area. It was time for the Tennessee Department of Transportation to take the action needed to protect what we had confirmed to be a remarkable discovery—remarkable not only for Tennessee but for the eastern United States as well.

4

The Public Announcement

As it happened, I had already taken the first step in what would lead to the eventual shutdown of the construction site by TDOT. On June 27, right after Larry Bolt informed me of Marta's discovery of the possible elephant bones, I sent an e-mail to my Nashville supervisor, Len Oliver.

After describing to Len the nature of the road project and where it was located, I told him about the fossil find. "Initial inspection of the clay material by Larry Bolt," I wrote, "identified the deposit as a possible ancient lake or large pond bed. Subsequent visits by myself and others in my office as well as geologists with TDEC, have disclosed a large assemblage of fossil vertebrate animals as well as plant debris (including leaves, berries and seeds)."

I went on to describe the layered clays and the presence of lignite before identifying our specific discoveries. "A number of fossils of interest have been found and identified from the site," I explained. "These include the following: tapir (herbivore about the size of a small cow), a possible giant ground sloth (a large grazing animal bigger than a horse), a turtle, and a mastodon (a relative of the modern elephant). Fossil bones include the skull, vertebrate, leg and feet bones and ribs of the above animals. . . . Our staff has also found pieces of tusks of the elephant-like mastodon. These bones and teeth and unidentified specimens are in our office here in Knoxville. Other fossil specimens are at the University of Tennessee McClung Museum for study and identification."

I noted the growing interest in the site not only by University of Tennessee scientists and state agency personnel but also by citizens of the surrounding communities. I closed my e-mail by suggesting that the TDOT Information Office be notified in case either the general public or the news media had questions.

Len promptly sent my message to Luanne Grandinetti, director of the TDOT Information Office in Nashville, and told her of my concerns. It was shortly thereafter that the media began to inquire about the goings-on at Gray.

On July 5, Kristen Hebestreet, a reporter for the *Johnson City Press,* contacted me. As I told her about the fossil find, I said that we were trying to get what bones we could out of the way of the road project. A few minutes later, Fred Brown of the *Knoxville News-Sentinel* called me. I repeated the same information I had given to the Johnson City reporter. Larry Bolt fielded questions as well, and the next day, July 6, the story broke in the papers.

In the *Johnson City Press,* Kristen Hebestreet's front-page story carried the headline: "Fossil find . . . Road project reveals treasure of ancient bones." The story began: "A state Department of Transportation geologist says 'the fossil find of a lifetime' has been unearthed in Washington County."

The *News-Sentinel* also carried its story on page 1 under the banner: "Major ET fossil find . . . TDOT geologists cite 'significance' of skeletal discoveries." Fred Brown wrote: "Tennessee Department of Transportation geologists have identified the bones of animals hundreds of thousands of years old in what is being described as a major fossil find in a road project near Johnson City."

The Associated Press picked up on the story as well, and newspapers such as the *Oak Ridger* in Oak Ridge, Tennessee, carried its account. Fred Brown filed an additional story on July 7, reporting in the *News-Sentinel* that TDOT had ordered the construction work shut down in the area of the fossil find and that a security guard would be posted there. Those of us who were worried about the potential destruction and unauthorized removal of the bones could now breathe more easily, at least for the time being.

After the first news stories broke, calls from newspaper and television reporters began flooding our TDOT Geotechnical Engineering office. The phones rang continuously on the morning of July 7. Unaccustomed to such attention and publicity, we were nervous as well as excited.

Reporter after reporter asked: "What have you found?" "What does it mean?" "What was it like at the time when these animals lived?" At this point, of course, we still had no precise handle on our discoveries. Although Rick Noseworthy's discovery of the alligator skull had cast some doubt on the Ice Age hypothesis, we still conjectured—for lack of better information—that the fossils were the remains of Pleistocene Epoch animals that had lived some 12,000 to 18,000 years ago.

With so many questions coming in, a group of us agreed to meet with reporters later that same day at the site. The contractor had even offered the use of its equipment to help excavate bones, although some of the workers were heard to complain: "What's all the fuss about?"

In addition, the Tennessee state attorney general's office had ruled on July 6 that State Archaeologist Nick Fielder would be responsible for the excavation, removal, and disposition of the fossil bones. This ruling came at the request of TDOT officials who had grown increasingly concerned about what to do with the bones and who was responsible for their safekeeping.

As Larry Bolt and I drove to Gray to meet Nick Fielder and the reporters during the late morning of July 7, we talked about what we would say. The attorney general's ruling especially concerned us. Neither Larry nor I knew Nick, and we were somewhat fearful that TDOT might be denied further involvement in the discovery and research at Gray.

Nick was at the fossil site when we got there, as were Larry Hathaway, Lanny Eggers, and Freddie Holly. Dr. Clark from the University of Tennessee was also on hand to meet Nick and learn more about the research possibilities. We gave Nick a quick tour of the site and described everything that had happened so far. We showed him where different animals had been found and what the lignitic clay looked like. We picked up several fossil bones as we walked around and also showed him plant remains that had been preserved as lignite.

Meeting Nick face to face eased some of my fears. He spoke with obvious authority, and his well-tanned skin was evidence of many hours spent in the field conducting archaeological investigations. It was clear that he appreciated the uniqueness of the site and the need to protect and study it. We were going to have to cooperate, no matter what happened. "I am going to need all the help I can get," he told us, "since I am getting into a field that is outside of my experience and expertise."

As an archaeologist, Nick was primarily familiar with human bones and artifacts, not with fossil animal bones. Larry and I both offered to help him.

Lanny Eggers, TDOT project administrator, discusses the activity at the Gray site with a television reporter from Johnson City, Tennessee.

Exploring the site, we found some freshly exposed tapir bones, including a partial jaw with a few teeth. The bones were a dark grayish brown, meaning that they had been exposed to the air for at least a few hours. Nick placed them on the ground to show the gathering reporters what they looked like and how they blended in with the clay. The arrival of reporters also drew a crowd of local residents, who were as curious as ever about the activity at the site.

The reporters represented newspapers and television and radio stations from the Tri-Cities area of Bristol, Kingsport, and Johnson City, as well as from Knoxville. They circled around the bones, pointing and talking. Photographers snapped away, while TV crews filmed. "I've looked at a lot of bones from archaeology sites," Nick told them. "I'm amazed at the excellent condition of the bones. The preservation of the bones is so good it's like the animals died recently."

Dr. Walter Klippel (left) of the UT Anthropology Department discusses the fossil finds with Larry Bolt (center) and George Danker of TDOT.

Each of us met with at least one representative of the media. At one point, Nick Fielder, Mike Clark, Larry Bolt, and I were giving interviews at the same time to different reporters. Dr. Clark waved his arms as he talked, while Larry pointed to different locations on the site. Nick and I both showed the tapir bones we had just found, explaining to the reporters what tapirs were and how they had disappeared from North America some 10,000 years ago.

The media treated our discoveries as a major news event. There were TV cameras mounted on tripods, reporters with note pads circulating through the crowd, and sound technicians testing their equipment. As exciting as all the media attention was, however, I knew that this exposure would forever change the fossil site. People would soon come from everywhere to see what we had found. Security would be needed, and access to the site would become difficult. Yet, I also knew that this attention was necessary to protect the fossils and to ensure future research.

During this media frenzy on July 7, we also did some work in getting the research started. Dr. Clark began that day with his analysis and mapping of the surface deposits. Wearing khaki shorts and a tee shirt with a safari hat, Dr. Clark was easy to spot on the site. His ability to focus on his work made him oblivious to the media activity going on around him.

To aid Dr. Clark in his research, Larry Hathaway, Larry Bolt, and I asked the contractor to begin excavating a series of terraced "benches" in the newly excavated roadway cut slope. Looking like the steps of a giant stairway, these benches were about five feet wide and four feet high and extended from the base of the cut slope to its top, a total of about twenty feet or more. Dr. Clark was thus able to examine and measure in minute detail the vertical sequence of the layers of clay and other deposits. This would help in determining their exact nature and plotting their geological history. Excavating the benches also led to the discovery of several new fossils: some more tapir remains that George Danker and I found, plus the discovery by Marta Adams of an unusual jawbone that would be identified a few weeks later. The richness of the fossil site seemed to have no limits.

After the barrage of television crews, print-media journalists, and photographers on July 7, we thought that the frenzy was over. However, it kept up for several weeks as newspaper and television reporters continued to announce details about the find. The following excerpts from area newspaper reports offer a good picture of the attention the site was receiving:

TENNESSEE'S TREASURE TROVE . . .
FOSSIL FIND EXHILARATES GEOLOGISTS

> The contractor thought the soft clay might not be right for road construction . . . so Larry Bolt, a geologist with the Tennessee Department of Transportation, had a track hoe dig into the soil. "I had no inkling what was going to show up," Bolt said Friday.
>
> —Reported by Jacques Billeaud,
> *Knoxville News Sentinel,* July 8

DEPUTIES AID IN EFFORT TO
PROTECT GRAY FOSSIL SITE

> Authorities are struggling to protect the Gray Fossil Site from souvenir seekers, some of whom are reportedly trying to sell the ancient remains at local flea markets.
>
> —Reported by Kristen Hebestreet,
> *Johnson City Press,* July 20

GRAY AREAS OF PREHISTORY . . . EPOCH FIND AT ROAD SITE UNEARTHS PORTAL TO CREATURES' PAST

Earlier this month, crews cutting through a connector road to the new State Route 75 in Gray, Tenn., made what could be a defining Pleistocene fossil find for the state.

—Reported by Fred Brown,
Knoxville News-Sentinel, July 23

ROAD CONSTRUCTION THREATENS FOSSIL SITE

State Archaeologist Nick Fielder crouches in the dark gray clay, gently raking dirt away from something hidden tens of thousands of years. He holds up and wipes the object, revealing a row of teeth from a tapir—a long nosed animal about the size of a small cow that fed on wetland grasses during the Ice Age. "What makes this unusual is that frequently at fossil sites, you find individual bones. Here we've found entire skeletons," Fielder said of the archaeological gold mine unearthed in May by workers widening state Highway 75.

—Reported by the Associated Press,
The Tennessean (Nashville), July 31

Needless to say, the media were feasting on the fossil discovery. This sort of public exposure was something to which we scientific types were unaccustomed. To us, the attention felt like something out of a Hollywood movie.

5

Research and Planning

I returned to the fossil site on July 10 to discuss research issues with Dr. Clark and Nick Fielder. Through Lanny Eggers, I had arranged to have the roadway centerline surveyed and staked back for location control on the site. This would aid future researchers by indicating the location of each fossil discovery with pinpoint accuracy. The surveying included the elevations of the terraced research benches that had been excavated a few days earlier. Thus, researchers would be able to say at what point above mean sea level each fossil was found.

This surveying data was a big help to Dr. Clark in his investigation of the geomorphology of the area—both in terms of what had already been uncovered and what would be found in the future. Beginning at the surface and working downward through the research benches and the cut slopes of the roadway construction, he was recording in precise detail each layer of soil that had been exposed.

Anthropology student Marta Adams (center) was instrumental in drawing scientific attention to the fossil deposit. At left is Dr. Mike Clark of the UT Geology Department, and Dr. Paul Parmalee of the UT McClung Museum stands at far right.

Martin Kohl, a TDEC geologist, examines some of the tapir bones he found during the summer of 2000.

Nick Fielder and I walked the site, looking at bones and discussing plans for the research area. We agreed that a team should be formed to define the scope of what needed to be accomplished. This would mean getting in touch with the various interested parties to arrange for an organizational meeting on site. So, after leaving Gray that day, Nick and I began working the phones, contacting mainly those who had been helping with the salvage operations that summer.

The meeting was set for the morning of July 13. Larry Bolt, George Danker, Rick Noseworthy and I drove to the site together that day. Nick Fielder met us there, and we agreed that he should lead the meeting, with those of us from TDOT assisting as needed.

The day before, Fred Corum, the regional engineering director for TDOT Region One in Knoxville, told me that I would be the department's contact person for the fossil site, dealing with personnel from TDOT and other state agencies. Anything involving the construction work would be Lanny Eggers's responsibility. I relayed this information to Nick Fielder.

Larry Bolt was the first to recognize that the blackish-gray clay uncovered at the S.R. 75 road project might be the deposits from an ancient lakebed or pond.

By 10 A.M. about twenty of us had gathered at the fossil site. The agencies and institutions represented were TDOT, the Tennessee Department of Environment and Conservation, the University of Tennessee, and East Tennessee State University.†

---

†Those who attended the organizational meeting—and were subsequently placed on the fossil research team—included the following: Nick Fielder, state archaeologist; Lanny Eggers, Freddie Holly, Larry Hathaway, Jamie Carden, Rick Noseworthy, Harry Moore, Larry Bolt, and George Danker, all with the Tennessee Department of Transportation; Bob Price, Tennessee Department of Environment and Conservation, Division of Geology; Mike Clark, University of Tennessee Department of Geological Sciences; Walter Klippel and Marta Adams, UT Department of Anthropology; Paul Parmalee and Gary Crites, UT McClung Museum; Larry Bristol, East Tennessee State University; Robbie Fleming, an ETSU student; and S. D. Dean, a private avocational paleontologist.

Above: Dr. Mike Clark, UT geology professor, examines a fossil as TDOT geologist Larry Bolt looks on.

Right: Rick Noseworthy, TDOT environmental coordinator, holds the alligator skull he found.

We talked at length about the future of Fulkerson Road. "It is a county road," Lanny observed, "and the County Commission of Washington County could close it if desired." Since a sizable area of the site was in the Fulkerson Road portion of the excavation, closing that road would allow for the fossil investigation to continue even after the S.R. 75 roadwork was completed.

Much of the discussion centered on the excavation of the fossil bones from the S.R. 75 roadbed. We first considered excavating five to ten feet of clay, then backfilling the excavation with shot rock up to the subgrade elevation. Under this plan, any fossils we were unable to salvage would remain buried beneath the road.

The use of shot rock was to be implemented along the entire section of S.R. 75 as it crossed the fossil site. I suggested that we employ a "geogrid" to help stabilize the roadbed. As I explained to the group, using a geogrid would require excavating the clay to a depth of only about three feet or so. A geogrid is made from a synthetic polymer and resembles a length of fencing material with grids of about two square inches. If the geogrid were placed atop the remaining soft clay and the shot rock were layered on top of that, the roadbed would have the necessary stability.

We also talked about what to call the site. Names that were considered—and discarded—included "Sulphur Springs," after the name of a nearby spring and the early name of the local community, and "Barry Site," in honor of the landowner who had donated the property for the building of Fulkerson Road years before. We finally opted for the name "Gray Fossil Site," after the present name of the community.

That same day, a crew of TDOT personnel conducted an auger drilling investigation of the site. Larry Bolt and David Barker, the TDOT engineer who had been part of the field inspection team that visited the site on June 8, monitored this activity. The purpose of the drilling was to determine the vertical and lateral extent of the fossil deposit. The procedure proved to be slow and tedious because the wet, sticky clay tended to clog up the drilling equipment—a problem similar to the one that had plagued the contractor's earth-moving machinery weeks before.

At this time, the goal was to keep the roadway in the same location that had been planned from the outset, regardless of whether Fulkerson Road was closed or not. This would mean extracting as many fossils as possible before the roadwork resumed. We expected to receive money from the Federal Highway Administration to fund the removal. TDOT had

The research included drilling into the sediments to determine the vertical and lateral extent of the deposit. Pictured, from left, are George Danker, David Barker, Larry Bolt, Jimmy McGill (back to camera), and Carl Elmore, all of the TDOT staff.

estimated that the drilling at the fossil site would be completed by July 21 and that a preliminary construction design using a geogrid would be available at that time or shortly thereafter.

Nick said that from this point on, there would be no further fossil collecting at the site by anyone until excavation plans were finalized. "All persons who excavate fossils in the future," he said, "will have to obtain a permit from my office in Nashville."

We agreed that a list of everyone attending the meeting should be given to the Washington County Sheriff's Department so that the guard on duty at the fossil site could screen those approaching the site. Only those persons on the list would be allowed to enter the newly restricted area.

Finally, Nick authorized the digging of two test pits to aid in planning the excavation procedure. This was to be carried out the following week. With the test pits, Nick wanted to see if fossils would turn up at depths of

The initial formation of the research team for the Gray Fossil Site brought together anthropologists, geologists, biologists, engineers, and university professors. The team included, from left, Bob Price, Larry Bristol, the author (kneeling), Jamie Carden, Dr. Paul Parmalee, George Danker, Dr. Walter Klippel, Rick Noseworthy, Larry Bolt, S. D. Dean, Nick Fielder, Lanny Eggers, Freddie Holly, Dr. Mike Clark, Marta Adams, and Robby Toole.

two and six feet below the current surface. If they were found at these depths, this would tend to support the assumption that we were dealing with an extensive fossil deposit. Nick also wanted Dr. Parmalee and Dr. Klippel to examine the contents of the trench excavations.

In the brief history of the newly named Gray Fossil Site, this meeting marked an important moment, and we posed for a group photo. We set our next meeting date for July 25 at 10 A.M., and again we would meet at the site.

On July 18, Larry Bolt and I returned to Gray to observe the drilling operations. I was also to meet with Luanne Grandinetti, who was visiting the site for the first time, and with Tim Whaley, a reporter with the *Kingsport Times News*. Luanne set up the meeting between the reporter and me after the newspaper requested the interview.

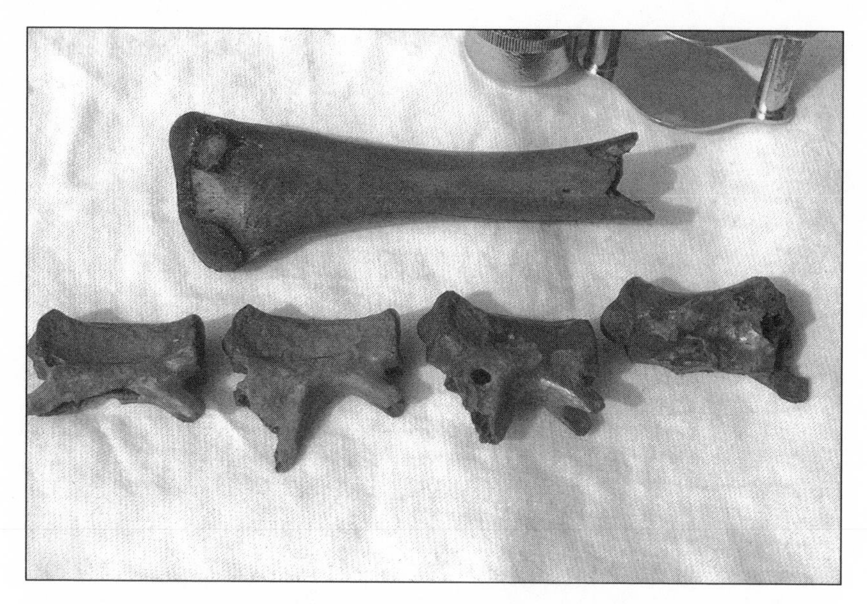

Additional excavation and research disclosed the presence of alligator vertebrae and a leg bone found by Marta Adams. These were the first such bones ever uncovered in Tennessee.

Accompanied by Freddie Holly, Luanne questioned me thoroughly about what we had found at the site and the circumstances of our discoveries. Tim Whaley had not yet arrived, and she wanted to make sure that she understood all the details. I could readily see that she was well suited to handle questions from the media.

We walked around the excavation area looking for anything significant. My eye was well trained by this point, and I noticed the tip of a fossil protruding from the clay. Pulling it out of the muck, I saw that it was a bone some five inches long, two inches wide, and about an inch thick. It was obviously the remnant of a large mammal. I called to the others to show them the find, and together we cleaned it off in a nearby pool of water. We had no clue as to what animal it came from, although a giant ground sloth or an elephant were our first guesses. As it turned out, it was from an animal that had never been found before in Tennessee. But that identification would not come until several weeks later.

When Tim Whaley arrived, Louanne, Freddie, and I showed him the find, and he seemed impressed. His colleague, news photographer Ned Jilton II, took pictures, and we discussed how I found the bone. It would make an interesting news article. Continuing our stroll, we soon found another unusual fossil—apparently the portion of a tooth or possibly an elephant tusk. It fragmented into many pieces as we dug it up.

Meanwhile, the drilling of the thick clay proceeded slowly. The drills penetrated depths of 70, 80, and over 100 feet. It often required more than an hour to drill a single hole. Several of the holes went to 128 feet, the greatest depth the drills could achieve, and still the black clay was coming out.

It now appeared that the clay deposit was oval in shape, about 600 feet wide by 700 feet long, and that the deepest parts were near the center. This suggested a possible sinkhole. Broken chunks of limestone and dolostone were mixed in with the clay, which indicated that we were dealing with a collapsed karst-type sinkhole.

If this was a sinkhole deposit, we thought, then it must have held a spring-fed pond—one that supported an environment replete with plants and animals, including crocodiles, turtles, frogs, and tapirs. Even the elephants would frequent the pond to drink and browse on juicy vegetation. Such ideas, of course, were no more than hypotheses at this point.

We wondered, too, what had happened to the animals. Their skeletons were largely intact. What had killed them, and how did they end up here? Drowning still seemed the likeliest possibility, but we hoped that the scientific research would eventually tell us for sure.

On July 20, I brought Bob Hatcher, Distinguished Professor in Geological Sciences at the University of Tennessee, Chris Whisner, a UT doctoral student in geology, and Jeff Muncey, a geologist for the Tennessee Valley Authority, to the site. Together they were working on the idea that there might be a deformation in the sediment due to possible earthquakes in the area in the distant past. They had no precise time frame in mind for the occurrence of such seismic events, but they hoped that if we could pinpoint the age of the fossils, this would help them to establish a date for the earthquakes.

Acting as the tour guide, I pointed out various aspects of the site's geologic character and extent. Chris had been to the site several times before with Peter Lemiszki, a geologist and head of the Knoxville Division of Geology office for the Department of Environment and Conservation. They had noted possible anomalies in the clay deposit that seemed to be evidence of long-ago seismic activity.

On this day they planned to measure "joint sets" in the sediments. They found several well-developed joints, or smooth fracture lines, in the blackish-gray clay along the new cut-slopes of Fulkerson Road. After measuring these joints, they showed me a possible soft sediment deformation called a "dewatering pipe," or "vent." Such vents were usually formed around cracks in the soil and sediment cover and provided an avenue for a mixture of groundwater and sand that might be pushed upward by a seismic vibration. Readily visible, the possible pipe structure was several inches in diameter and extended for more than thirty feet up the slope.

The pipe structure contained numerous clay fragments that were oriented at different angles all up and down the "pipe." It was actually possible to see the curved layers of clay parallel to the walls of the pipe structure. This appeared to be good evidence to support their earthquake hypothesis.

Dewatering pipes occur when seismic vibrations shake saturated sediments, causing the material to liquefy. As a result, the liquefied silt and sand spurt up toward the surface along the sediment cracks. This upward thrusting is due to the overburden pressure caused by the overlying sediments. The pipe structures often result in "sand blows" at the surface where the liquid material sprays out into the open air. Occurrences of this sort were common during the 1811 and 1812 earthquakes in West Tennessee. Usually called the New Madrid Earthquakes, those events formed Reelfoot Lake in the northwest corner of the state.

Dr. Hatcher told me that he, Chris, and Jeff were preparing a paper for the annual meeting of the Geological Society of America, scheduled that fall in Nevada, and they planned to use the information they obtained at the Gray Fossil Site in their presentation. Ancient soft-sediment deformations of the sort seen at Gray are rarely encountered. It was now apparent, thanks to the work of Dr. Hatcher and his colleagues, that the fossils were not the only unusual features of the site; the deformed clays were themselves worthy of study.

Others were busy at the site that same day. The UT Department of Anthropology had people there digging the two test pits that Nick Fielder had requested. Each pit was about four feet wide by eight feet long. One test pit was approximately one and a half feet deep, and the other was six feet deep. There were fossil bones throughout both of them, thus confirming Nick's suspicions. Several anthropology students sifted through the soil as it was removed, placing the bones in plastic bags and marking the bags with identification numbers for later research.

At the bottom of the six-foot pit was a complete tapir skeleton. Dr. Parmalee and Dr. Klippel were down in the pit, carefully removing each bone. Before I left that day, they had retrieved much of the remains, which they planned to take back to the university for additional study. Again, I thought, it was another successful day at the Gray Fossil Site.

As scheduled, the second meeting of the research team was held on July 25. About twenty-five people were in attendance, including Rab Summers of Summers Taylor, Inc. Much of the discussion centered on this plan: S.R. 75 would *not* be rerouted, and any undiscovered fossils lying directly beneath the road would be left there for indefinite preservation. Nick summarized the findings in the two test pits dug on July 20, both of which had revealed fossils at depths close to the proposed subgrade elevation.

Nick also described a proposal for the establishment of a study area located mainly at the southeast corner of the new Fulkerson Road–S.R. 75 intersection. University researchers, he said, believed that closing Fulkerson Road might be a good idea. Under this plan, the road would be closed and converted to a cul-de-sac, which would then be used as a parking and staging area for continued fossil excavation work. Nick said that he would contact the Washington County Highway Commissioner and other local officials to determine whether the cul-de-sac proposal posed any problems. He also suggested that the study area could be subdivided into a number of smaller zones to facilitate research.

As the meeting closed, Nick informed us that the Tennessee Division of Geology would map the site geology. He also announced that I, representing the TDOT Geotechnical Engineering Section, would act as his second-in-command on issues related to the Gray Fossil Site.

6

The Politicians Come

The media reports about the Gray Fossil Site did not escape the attention of state officials in Nashville, and by late July Gov. Don Sundquist was making plans to visit Gray to see firsthand what all the commotion was about. Phone calls about the fossil finds had been pouring into the governor's office, as well as those of various state representatives and senators. The governor was set to arrive on August 7, joined by J. Bruce Saltsman, the commissioner of transportation, and Milton H. Hamilton Jr., commissioner of environment and conservation.

Many citizens of Gray and other nearby communities were as excited about the fossil discoveries as those of us who had been studying the site were. Undeterred by the blazing midsummer sun, a crowd of onlookers would arrive at the site each day and watch us as we excavated bones. Now that guards had barred unauthorized personnel from entering the actual construction area, the sightseers would gather along the existing highway and monitor the salvage and research activity, some of them using binoculars.

There was now growing public sentiment that the bones should stay in the vicinity of Gray rather than be taken to the University of Tennessee in Knoxville. In a July 11 story that appeared in the *Johnson City Press,* Robert Houk reported: "State and county officials said Monday they have been bombarded by calls in the past week from Washington County residents who believe the fossils unearthed at a highway construction site in

Sulphur Springs should remain in the area." Ten days later, Houk reported that Governor Sundquist had promised a state legislator from the area that he would "consider building a state museum in Washington County to display the fossils unearthed in Gray." Clearly, some questions had been raised that the governor and other state officials would soon have to address.

On August 2, I met Nick Fielder and Larry Hathaway at the site to discuss the governor's upcoming visit. We strolled around the excavation and identified areas that we thought would be important for the governor to see. How to steer the tour group around the muddy portions of the deposit was one of our concerns.

It was on this walk that yet another significant fossil turned up. Near the proposed new intersection of S.R. 75 and Fulkerson Road, an object protruding from the ground caught my eye. From what I could see at first, it seemed to be no larger than a penny. I started digging around the piece and noticed that it had an enamel-like covering. After several minutes of careful digging with a knife, Larry Hathaway and I unearthed the fossil. It appeared to be a portion of a tusk, formed in longitudinal layers, with a crosshatched pattern visible at one end. It was gray-black in color and had a smooth, shiny surface. It was in much better condition than the section of tusk I had found on June 26. That one was heavily weathered and looked like a piece of yellowish-brown wood.

The new fossil measured about two and a half inches wide by about three and a half inches long; it was about an inch thick. It was shaped in a broad arc on one side and somewhat flat on the other. Though sure that additional remains would be found at this spot, we decided to leave them in the ground for excavation at a later date. I took the piece I had unearthed back to the office for safekeeping.

On August 7, I accompanied Fred Corum to the Tri-Cities Airport to await the governor's plane. We first met with the Tennessee Highway Patrol officers who would transport the governor and others to and from the Gray Fossil Site. They were in plain clothes—suits and ties with concealed shoulder holsters.

The governor's plane arrived at around 9:30 A.M. The scene looked like something from a movie set, with the plane landing and the Highway Patrolmen rushing onto the tarmac to meet the governor and those traveling with him. Fred Corum and I waited at the terminal door. Disembarking from the plane were Governor Sundquist, his daughter Tania Williamson, Commissioners Saltsman and Hamilton, Nick Fielder, several members of the governor's staff, and a security detail consisting of Tennessee Bureau of Investigation officers.

The state troopers quickly and determinedly took us to the fossil site. We rode in three sedans and traveled down State Route 75 from the airport, passing across the interchange of Interstate 181 (now Interstate 26) and continuing down S.R. 75 to Shadden Road, where we turned left. We were approaching the site from the Fulkerson Road side. This would bring us to a large graveled area that the construction crew had prepared in order to accommodate the large number of vehicles that were expected that day.

As we turned right onto Fulkerson Road, we passed two beautiful old homes that appeared to have been there for a hundred years or more. We learned that these were the residences of the original owners of the property that had become the Gray Fossil Site.

After traveling about a half-mile, we arrived at the site, where a large crowd of media people and politicians met us. A number of the local citizens were also there, waiting to hear what the governor had to say about the fossil discoveries.

As we left our cars, the reporters and others immediately descended upon the governor. The Highway Patrol officers had to help keep the crowd at arm's length so that we could proceed with our tour of the site. The governor's entourage regrouped and started toward the main construction area to a spot that overlooked most of the site. From this vantage point, the governor could see the area where the clay was exposed and where a large number of the fossils had been found.

Off to one side, I heard some plain-clothes officers yell, "Hey, you up there in the woods! Come out where we can see you. NOW!" There was no response, so the officers immediately went into a defensive stance, with one man drawing his firearm. The order was repeated, this time more loudly.

The bushes and tree limbs swayed, and there was mumbling in the woods. A cry came out: "Don't shoot. We're surveyors. We work with TDOT." Once the officers could see the four men with surveying equipment, the excitement was over. Governor Sundquist seemed not to have noticed the commotion. Surrounded by reporters, he was otherwise occupied.

Nick Fielder led the group around the site, pointing out the unusual clay deposit and the different spots where fossil bones were found. He picked up some pieces of lignite and showed them to the governor. He explained that I had found some of the bones and that I had them with

State Archaeologist Nick Fielder (left) talks about the fossil site conditions with visiting officials, including Gov. Don Sundquist (in sunglasses). Photo by Larry Hathaway.

The author shows a portion of a tapir skull to Gov. Don Sundquist.
Photo by Larry Hathaway.

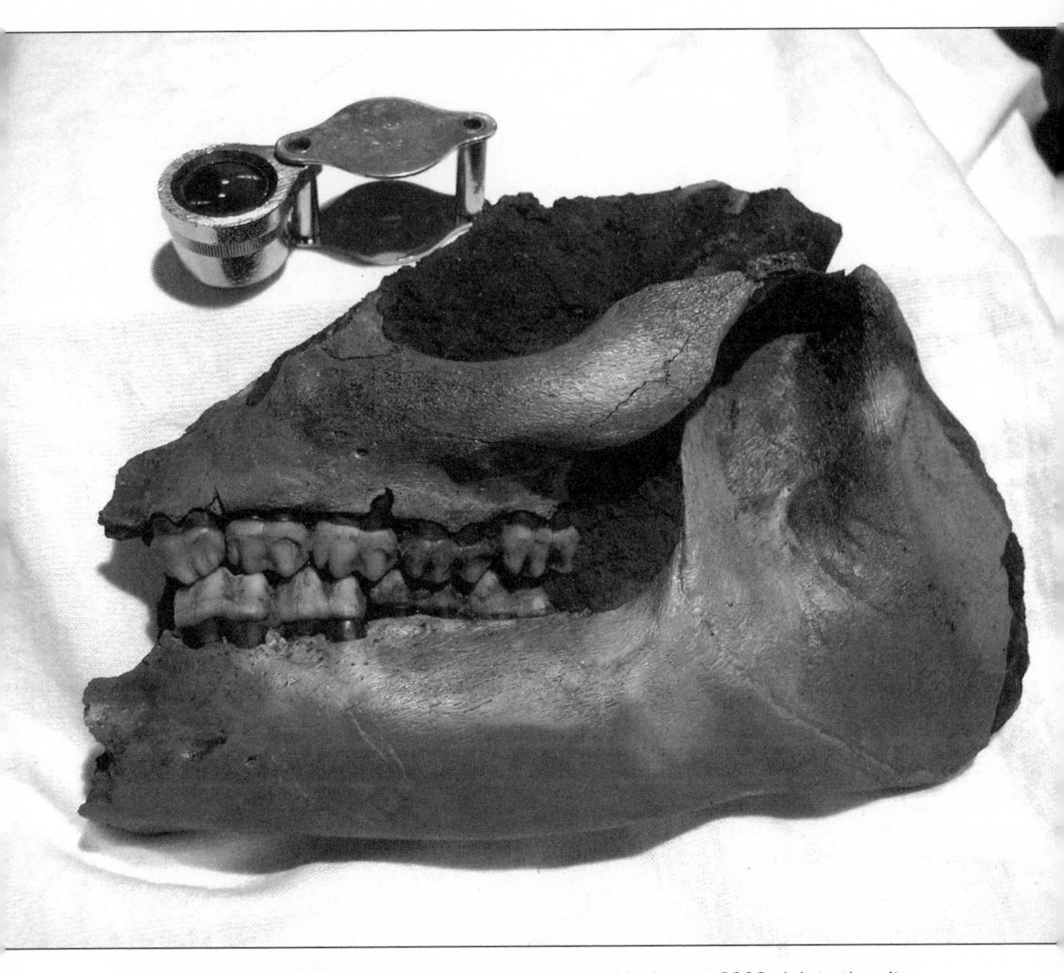

The tapir skull that was shown to the governor during his August 2000 visit to the site.

me. So, one by one, I removed fossil bones from a box I was carrying and handed them to Governor Sundquist.

The reporters crowded in, and a hundred camera shutters clicked. First, I showed the governor the half-skull of a tapir, about the size of a frying pan, and pointed out its features, including its beautifully preserved teeth. All around us people gasped. It was both exciting and gratifying to hear their responses.

Next, I took out the "big tooth" that I had found on my June 26 visit to the site with Don Byerly. This, I explained, was thought to be the tooth of a giant ground sloth, probably from the Ice Age. The governor held it up for all to see. "This is unbelievable" and "Wonderful" were phrases I kept hearing from the crowd.

Next, I unveiled a femur from a tapir and passed it around. It was about nine to ten inches long and a chocolate color. It contained the ball-joint portion where it had attached to the hip or pelvis.

Six tapir teeth, which I removed from an envelope, came next. These were about an inch and a half long and black except for the medium gray enameled portion at the tip of each tooth. Once part of the tapir's mandible, or lower jaw, these teeth had been found loose amid the disturbed clay.

Finally, I displayed a portion of an elephant's tusk. It was about four inches in diameter and six inches long and weighed about five pounds. This particular piece had been discovered by Christy Brown, a TDOT design engineer from Knoxville. It resembled a piece of wood with a white crusty and lineated surface.

The display of fossils completely "wowed" the governor and the various onlookers. Impressed by the need to preserve the bones, Governor Sundquist asked Commissioner Saltsman to have the TDOT staff review the construction project with all the interested parties. He wanted us to consider alternatives to the present proposed roadway that cut directly through the fossil deposit.

The governor suggested that TDOT should look at three possibilities in its efforts to resolve this situation: (1) keeping the road in its present location and removing as many fossils as possible before completing the road construction; (2) building a bridge over the deposit in the same location as the present road construction; and (3) moving or relocating the road completely away from the fossil site.

It was interesting to note that the governor's daughter, Ms. Williamson, was quite excited about the findings at the fossil site. I escorted her to an area where I thought there might be some readily visible fossils so that she could try her hand at bone hunting. She located several specimens and called her father over to show him what she had found. It was my impression that her interest and enthusiasm rubbed off on the governor, who also bent down and started looking for bones. When he turned up several small pieces, he was impressed still further by the extent of the deposit.

Within an hour or so after it began, the meeting was over and the governor and his entourage departed for the airport. The media people packed up their equipment and soon left the site.

On the next day, August 8, articles in the *Johnson City Press, Elizabethton Star,* and *Kingsport Times-News* all reported that Governor Sundquist wanted the fossils to stay in the northeast Tennessee area. Kristen Hebestreet's story in the *Press* noted the governor's belief that the fossils "could be a tremendous boost to tourism here." In the *Times-News,* meanwhile, Tim Whaley reported that the governor said "the state probably has enough spare change to purchase extra property and implement design changes to the State Route 75 road project that uncovered the fossil find."

These news reports clearly reflected the concerns of local residents about the fossil site. The citizens of the surrounding communities obviously felt that the fossils belonged to their area, that they should stay there, and that their communities should be the ones to benefit from the discovery. The economic rewards that might come with increased tourism were one potential benefit, but there would also be educational and research benefits to local schools and to East Tennessee State University in nearby Johnson City. Thus, the disposition of the fossil bones had become a political issue that Governor Sundquist would ultimately have to resolve.

The governor's request for alternative ways of dealing with the site was the subject of the next research team meeting on August 16. We were charged with reviewing and debating the alternatives and making a recommendation to Commissioner Saltsman. It was another in a string of hot days, and by 10 A.M. everyone was perspiring. The humidity was high, and the haze was thickening.

We had gathered in the middle of the site and were using the hood of one car and the trunk lid of another as a makeshift table. We spread out our roadway plans, drilling information, and geologic mapping data, using a few fossil bones as paperweights.

Jim Ziegler, TDOT assistant to the state engineer, conducted the meeting. Representatives from numerous TDOT offices—Design, Structures, Geotechnical, Survey, Construction, and Safety—were present, as were representatives from other state agencies and the University of Tennessee.

Huddling around the plans and data reports, we talked about the three alternatives that Governor Sundquist had suggested. Alternative A—keeping the road in the same place—was essentially the plan we had first discussed on July 13; it entailed the use of a geogrid to stabilize the roadbed. Alternative B—building a bridge over the site—was the most expensive and generated considerable debate. Costing an estimated $1.4 million, it

TDOT engineers discuss possible alternative roadway designs that would protect the fossil deposit. Note the bones on the car hood. Gathered, from left, are Andrea Hall, roadway designer; Mike Agnew; Pete Faulkenberg; Jim Zigler; and Robert Garrett, with his back to the camera.

required drilling and construction that would greatly disturb various portions of the fossil site.

Alternative C was the overwhelming preference since it would leave the site undisturbed and available for continued research. We reviewed a proposed alignment layout that portrayed a new roadway location to the west and north of the existing roadway, well outside the area where the fossils were being discovered. The estimated cost of relocating the road was between $700,000 and $800,000, and it would require the removal of two houses and a barn.

We put the three alternatives to a non-binding hand vote, and, to no one's surprise, we agreed to recommend Alternative C. The results of our discussion were subsequently presented to Commissioner Saltsman, who in turn relayed the information to the governor.

By August 22, when I next returned to Gray, the erection of a chain-link protection fence around the fossil site was well underway. While construction of the roadway had stopped in the vicinity of the site, it continued on both sides of it. It had rained recently, causing some erosion of the excavated area. Larry Hathaway and I strolled around the site, hoping to see whether any significant new fossils had been exposed, but we found only a few bone fragments.

The humidity was so oppressive that day that it seemed to hang on the trees. It made me think of the time when the prehistoric animals whose remains we were now discovering had roamed the landscape foraging for food and water. This ground felt like a special place where the spirits of the animals hovered and watched all our digging and discovery effort. If only these animals could have known what an impact they would have these many years later. Their bones spoke of a time that we would continue to discover and explore.

On August 25, I received a call from Paul Parmalee about the jawbone and big tooth that I had found on June 26 and had later shown to Governor Sundquist. Dr. Parmalee had contacted the Illinois State Museum, where he once had worked, to inquire about the possible origins of the tooth.

Near the end of July, Marta Adams had found a remarkably similar tooth-and-jaw fossil in the excavated "benches" that Mike Clark was studying along Fulkerson Road. When she and Dr. Parmalee came by the office one day and showed it to me, we all noted its similarities to my discovery, surmising that it had come from the same kind of animal. Grayish green to black in color, Marta's specimen showed no sign of weathering. The tooth and jaw that I had found, on the other hand, was a yellowish-tan color and did appear to be weathered.

Dr. Parmalee later told me by phone that he had received word from the Illinois State Museum that the tooth might be that of a rhinoceros, possible early Pliocene or late Miocene in age. They advised him to contact Michael R. Voorhies, a professor of vertebrate paleontology at the University of Nebraska in Lincoln.

The tooth-and-jawbone fossils that Marta and I had found separately, along with toe and leg bones that Martin Kohl had found, were packaged and mailed to Dr. Voorhies for precise identification. We were expecting his answer shortly.

Rhinoceros! What on earth, we asked each other, was a rhino fossil doing in Tennessee in what we supposed were Pleistocene Ice Age deposits? Rick Noseworthy's alligator skull had already aroused our suspicions that the sediments might be older than we previously thought, and this latest news upended our suppositions even more. What, we now wondered, did this mean for reconstructing an ancient environmental model of the area? If the rhino speculations turned out to be true, the Gray Fossil Site would be a discovery of international significance. The implications were numerous and astounding. At our office, we discussed the issue constantly for the next several days, while still trying to conduct our TDOT business as usual. We also talked to Dr. Parmalee and Marta Adams about the possibilities of this major discovery.

On September 12, Dr. Parmalee called me with news about an additional piece to the puzzle. Although he had not yet heard from Dr. Voorhies at Nebraska, he had received word from Alan Holman, a professor of paleontology at Michigan State University, about the turtle shells that had been found at the site. Dr. Holman's research had tentatively identified them as fossils of water turtles, similar to the kind often seen in old ponds or swamps, usually on or around partially submerged logs. Such turtles are popularly known as "scooters," "sliders," or "skimmers." Dr. Holman said that the shell carapace appeared to be different from other species in their collection; it might even be a new species.

The true age of the fossils was now an ever-growing mystery. "Red flags" that pointed to something other than an Ice Age deposit were popping up all over.

7

The Final Decision

Governor Sundquist and Commissioner Saltsman came back to Gray on September 15 for a press conference. It was time to announce the decision that would decide the fate of the fossil site and the road construction project. Louanne Grandinetti, TDOT information director, accompanied the governor and commissioner.

Once more the media were there in force. Camera crews, reporters from television stations and newspapers, photographers, other politicians, local citizens, and a few TDOT personnel were all on hand, as were the security personnel who were there to insure the governor's safety.

The time was 2:30 P.M. Governor Sundquist told the assembled crowd that TDOT would move the S.R. 75 project to the west by approximately five hundred to six hundred feet. This new alignment would completely bypass the fossil deposit, thus saving it for scientific research and education. Commissioner Saltsman then described what the relocation of the road would entail in terms of design and construction, noting the estimated cost of $700,000 to $800,000.

This was big news. As the cameras clicked, the reporters fired their questions: "How long would it take?" "Where would the road actually be located?" "How many houses would be taken for the relocation?" "Who would own the fossil site?" "Who would be in charge of the fossil site?"

Definite answers were not available at the time, however. Commissioner Saltsman assured the reporters that the appropriate course of action

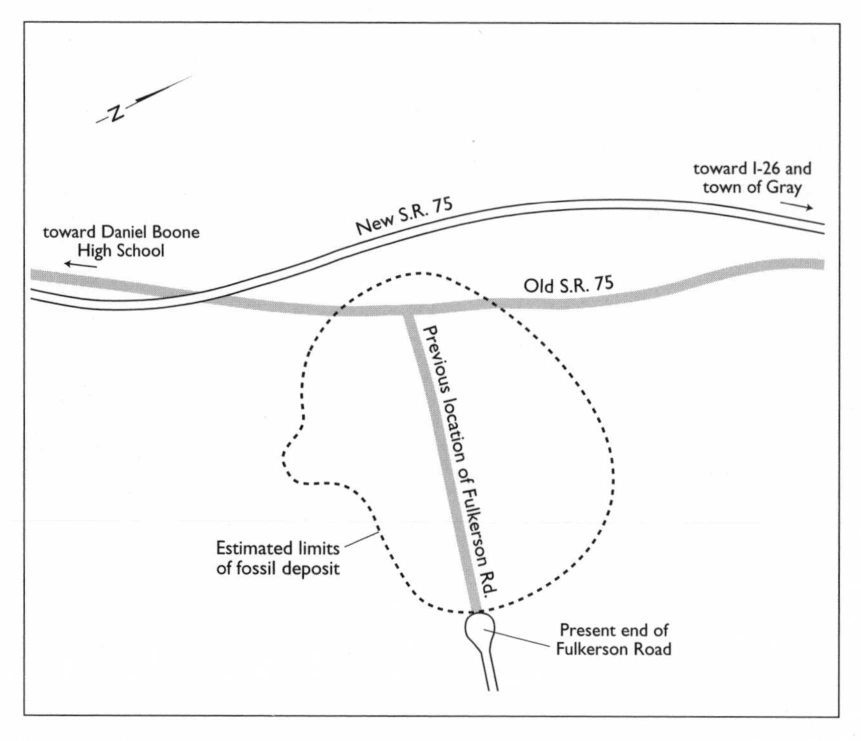

Location of Gray Fossil Site in relation to old and new locations of S.R. 75.

would be taken for designing and constructing the relocated roadway section. The TDOT Geotechnical Engineering Section, he said, would drill in the area of the proposed relocation to make sure that the new ground contained no fossil deposits.

As Commissioner Saltsman explained, the construction plans would be amended to account for the fossil site bypass, and the roadway construction would then be completed.

This was a major announcement on behalf of scientific research and education. That a political decision was made in favor of science in Tennessee set a historic precedent. Interestingly, I learned that day from Nick Fielder that Governor Sundquist had taken a geology course in college that had introduced him to fossils. The course left him with an enduring interest in the geosciences and paleontology. Thus, when he heard that an unusual fossil deposit had been discovered that might be of national and international importance, he had wanted to see the site firsthand and consider ways of preserving it. I have no doubt that his acquaintance with

In September 2000, Gov. Don Sundquist (center) announced that the road construction at Gray would bypass the fossil site. To the right of the governor is Transportation Commissioner Bruce Saltsman.

geology was what prepared him to make the decision that saved the Gray Fossil Site for posterity. With a shudder, I thought about what might have happened if the Tennessee governor had had no interest in geology or fossils. We might well have lost one of the major scientific finds of the last fifty years.

On September 18, another big announcement came. Michael Voorhies and his colleagues at the University of Nebraska, Lincoln, had positively identified the bones that Dr. Parmalee and Dr. Klippel had sent him as having come from a rhinoceros.

Dr. Voorhies was stunned that such bones had turned up in East Tennessee. He identified the tooth-and-jaw pieces as well as the carpals and leg bones as that of *Teleoceras*—a stocky creature that looked more like a hippopotamus than a modern-day rhino. This was the same genus of rhino he had studied at the famous Nebraska Ash Fall site where numerous Miocene Age rhinos were engulfed by volcanic ash. (We later determined that the bone I had found on July 18, the day I had first met Luanne

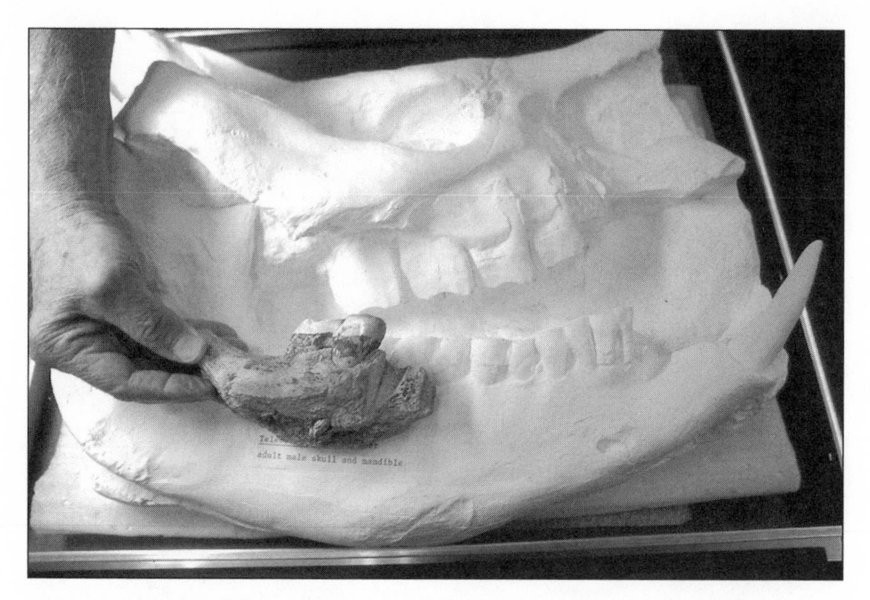

The rhinoceros tooth that the author found is compared to a plaster cast of a skull of the same species, Teleoceras, which was found in Nebraska by Dr. Michael Voorhies of the University of Nebraska.

Grandinetti at the site, was also from a rhino. We made this identification by comparing my bone to one Martin Kohl had found.)

The verification of the rhino bones clinched it. The fossil site dated from the Miocene Age, which meant that they were at least 4.5 million years old and quite possibly up to 18 million years old. Future research may pinpoint the exact age of the site, based on fossils that have yet to be unearthed.

To the geologic community in Tennessee, this was a landmark revelation. Never before had a Miocene Age deposit been found in the state. And not only that, but it also contained large vertebrate animals, including species previously unrecorded in the area.

The headline in the *Knoxville News-Sentinel* on September 19 read: "Ancient rhino is latest find . . . Creature might have lived as much as 18 million years ago." Fred Brown's story began: "University of Tennessee scientists working with paleontologists from other universities have positively identified several bones found at the Gray Fossil Site as those of a small rhinoceros that lived millions of years ago in East Tennessee."

Truly, there was much to celebrate, but there was also much work to do to protect and encourage scientific research and educational opportunities for the public.

The Miocene Epoch and the Gray Site

A subdivision of the Tertiary Period of the Cenozoic Era, the Miocene Epoch began about 24 million years ago. The bones found at the Gray Fossil Site are now believed to be of late Miocene age, or about 5 million years old.

"Miocene" is a term derived from the Greek words "meion" (less) and "ceno" (new). Similarly, the Pliocene Epoch, which follows the Miocene, gets its name from the Greek words "pleion" (more) and "ceno" (new). The British geologist Sir Charles Lyell (1797–1875), who studied fossils in the strata of the Paris Basin during the 1820s and 1830s, is credited with coining these names. He found that the rocks in the uppermost, or Pliocene, layers of the Paris Basin contained a higher percentage (90 percent, in fact) of fossil remains of living mollusk species, and thus were "more new," whereas the underlying Miocene layers contained very few fossils of living mollusk species (only 18 percent of the total fossil deposit), and thus were "less new."

The Miocene is thought to be a time characterized by a relatively warm climate and widespread grasslands, which replaced much of the forested land that had earlier prevailed. Grazing animals—those feeding on grasses—competed successfully against browsing animals, which feed on leaves, twigs, branches, and shrubs. Horses and camels evolved rapidly and flourished. Rhinos, giraffes, hyenas, tapirs, and the early elephant-like animals, such as mastodons, also thrived among the grasslands and savannas around the world.

In North America, most fossil remains from the Miocene Epoch are found in the western regions, such as the Great Plains. Records of Miocene terrestrial life are rare east of the Mississippi River and, prior to the discovery of the Gray Fossil Site, were found mainly in northern Florida at the Thomas Farm Fossil Site and at a location in north-central Indiana called the Pipe Creek Sinkhole Site.

The magnificent assemblage of vertebrate animal remains near Gray, Tennessee, has provided scientists with the opportunity to discern valuable information about the animal and plant life in the interior sections of eastern North America. We now know that browsers and grazers inhabited the East Tennessee region, suggesting a warm climate and a landscape with grasslands and savannas, as well as forests. The discovery of rhino and elephant fossils, along with those of aquatic vertebrates, such as fresh-water turtles, frogs, alligators, and fish, offers ample support for the warm climate theory for the Gray Site.

A particularly notable feature of the Gray Fossil Site is its large number of articulated tapir skeletons. Most fossil tapir sites yield few specimens, mainly teeth and leg bones. At Gray, however, extensive remains of some twelve to fifteen individual tapirs have been recovered, including juveniles as well as adults.

As climatic conditions changed and the Miocene Epoch gave way to the Pliocene, some animals adapted while others became extinct. Tapirs, for example, persisted in North America for another several million years, until the close of the Ice Age. Today, tapir species are still found in South and Central America, as well as in Sumatra and the Malay Peninsula. The reign of the rhinoceros in North America, however, ended with the Miocene Epoch, and today rhino species are found only in Asia and Africa. ■

18

Afterwards

For the next several months, the University of Tennessee scientists, including Paul Parmalee, Walter Klippel, and Mike Clark, met with various members of the TDOT staff, including me, in order to prepare for and organize the fieldwork that would begin soon. The ad hoc research team that had been established to determine what to do with the site had now become the group that would decide how to study the site. A political fight was beginning to take shape.

Word was spreading through the scientific community that the site was, in all likelihood, of Miocene origin. Finding fossils of this age in an area where none had previously been recorded would certainly help in gaining research support for the site.

The first order of business was to obtain funds to support the work at the site. This would prove to be a formidable task with many ups and downs, and it quickly emerged as the number-one problem. With the state in a fiscal crisis, the Tennessee legislature was engaged in a debate over whether to adopt an income tax or other revenue-producing measures that would keep state government services operating.

At the close of the 2000 calendar year and during the early months of 2001, it became increasingly apparent that government funding of the needed research for the Gray Fossil Site would be very difficult to obtain. The dilemma for the researchers was that large funding organizations such as the National Geographic Society and the National Science Foundation base their funding selections on documented scientific findings. At this time there had not been any extensive scientific research done on the Gray Fossil Site discoveries. Our work to date could mainly be described as a salvage effort.

The scientists who had been working at the site asked TDOT and TDEC for funds to investigate the fossil deposit, its origins, and content. Their concerns prompted Nick Fielder to call another meeting of the Gray Fossil Site research committee on April 9, 2001. Gathering at the Region One TDOT office complex in Knoxville, we learned that possible funds from TDOT and/or TDEC might indeed be forthcoming. The prospect of having funds available for the start of the 2001 summer fieldwork season was exciting, and we divided the research to be done into several distinct fields of study: geophysics, large vertebrate and small vertebrate animals, plants, clays, stratigraphy (a branch of geology dealing with the composition and distribution of the earth's strata), geomorphology, and seismology. Different members of the committee volunteered to oversee those aspects of research that were in their fields of specialization. Mike Clark, for instance, was to oversee the geomorphology research.

Over the next several months, however, those who wanted to do the research at the site, such as Paul Parmalee, Walter Klippel, and Mike Clark, began to feel uneasy about the prospects for funding and also about their continued access to the site. The State of Tennessee began to drag its feet over allocating funds, and it was also unclear who would be given responsibility for protecting the fossil bones that had already been discovered and retrieved. Although the University of Tennessee in Knoxville had the expertise and facilities to do the research, local sentiment for keeping the bones in the area of their discovery eventually dictated that the primary responsibilities for research should be placed in the hands of East Tennessee State University in Johnson City.

Since ETSU had no specialist in vertebrate paleontology on its faculty, Governor Sundquist decided in the spring of 2001 to authorize funding for just such a position. ETSU officials began advertising for the position in May by placing advertisements in the appropriate journals and by contacting various universities around the country. By the end of the summer, ETSU had hired Steven Wallace, who had just completed his Ph.D. at the University of Iowa. Dr. Wallace specialized in small mammals of the Ice Age while studying at Iowa, and he was particularly interested in the Miocene Epoch.

As it happened, I had met Dr. Wallace, who prefers to be called "Wally," that July when he was being interviewed for the job at ETSU. Larry Bristol, a geology professor at ETSU, was showing him the fossil site when I happened to stop there en route to Mountain City in Johnson County while on a personal trip with my wife. When I asked Dr. Wallace what he thought of the fossil deposit, he told me, "This is a remarkable site. Whoever ends up here will have their career waiting on them."

A man of slender build with a big smile, Dr Wallace seemed eager that day to begin combing through the hidden treasures of the fossil site. Somehow I felt that he would be the scientist chosen to direct further research at Gray.

Dr. Wallace was on site by September 2001 and began developing his research plans. He discussed his ideas with both Dr. Parmalee and me. I met "Wally" on several occasions that autumn to show him where certain fossils had been found. These included not only the numerous spots where we had unearthed tapir bones but also the locations where I had found the rhinoceros jawbone and tooth and the elephant tusks. Along with Dr. Parmalee and Marta Adams, I also briefed him on how the site had been discovered, on our salvage efforts, and on the planned relocation of State Route 75.

On October 9, the Tennessee Department of Transportation erected a new sign at the excavation. It featured white letters on a brown background, and its wording was simple: "Gray Fossil Site." Next to those words was an illustration of a Miocene Epoch rhinoceros skull, genus *Teleoceras.*

In December, TDOT requested bids for the construction project that would relocate S.R. 75 near the fossil site. Baker Construction Services, Inc., a company based in Bluff City in Sullivan County, submitted the winning bid of about $969,000. Work began in early March 2002 and was completed by the late summer of 2003.

Meanwhile, several enclosures were placed on the fossil site that would permit year-round fossil excavation, even in inclement weather. TDOT employed four greenhouse-type structures that had previously been used for an archaeological excavation on a road project in the town of Townsend in Blount County. These were transported to Gray and re-erected at the fossil site, which was now fully bounded by a chain-link fence, during the fall of 2001 and winter of 2002. Each enclosure measured about fifteen feet wide by fifty feet long.

The Tennessee Department of Transportation put up a new sign for the fossil site. The skull depicted on the sign is that of the extinct rhinoceros *Teleoceras*.

During the erection process in late November 2001, Dr. Wallace had noticed some small bones exposed in the clay close to where one of the greenhouses was being constructed. With this chance discovery, Dr. Wallace got his first taste of the excitement that so many of us had experienced during the summer of 2000.

As he picked away at the clay, he soon realized that he had found the toe bones of a tapir and that they were still connected to the lower leg bones. He continued the excavation until he reached the pelvis area. He had uncovered the back portion of a tapir skeleton. A few days later, while strolling in the same general area, Dr. Wallace discovered a second tapir skeleton in much the same way that he had found the first.

Dr. Wallace's elation with his finds was evident when he later filled me in on the details. "Welcome to the Gray Fossil Site!" I told him.

In the spring of 2002, Dr. Wallace began laying the groundwork that would prepare him and other scientists to do research at the site. Using

Dr. Steven Wallace and his students at East Tennessee State University found this beautifully preserved turtle shell in the spring of 2002.

TDOT's previous survey work as a guide, Dr. Wallace developed a grid of geographic survey points with which to reference all future fossil finds at the site. The grid was based on a north-south/east-west orientation using iron-pin concrete markers previously set by TDOT engineers.

That June, Dr. Wallace began a true scientific excavation process using students enrolled in ETSU's new summer paleontology class. Each time the student excavators discovered a bone, a precise measurement of its location was made and entered into a computer database. The June 2002 excavations turned up a new turtle species along with several new tapir bones.

As of this writing, the scientific excavations have just barely scratched the surface of the clay deposits. Dr. Wallace told me that he expects the site to eventually gain international recognition for the large number of tapir fossils found there. He anticipates that future discoveries will further enhance the character and distinction of the Gray Fossil Site. (And, indeed, in February 2004, just as this book was going into production, yet another remarkable find—that of a red panda tooth—was announced in the press.) Current funding for research has come from the State of Tennessee, but grant requests have been submitted to the National Science Foundation, the National Geographic Society, and the Discovery Channel to raise additional funds.

On September 26, 2002, Governor Sundquist announced that he had awarded an $8 million Federal Highway Administration ISTEA grant to ETSU for the building of a welcome center–museum and educational facility at the Gray Fossil Site. The stipulations of the grant required ETSU to match the $8 million with $2 million for a total of $10 million. Construction was to begin in the late spring of 2003, with the completed facility scheduled to open to the public in 2005.

In the early winter of 2002, Dr. Parmalee and Dr. Klippel, along with Peter A. Meylan of Eckerd College and J. Alan Holman of Michigan State University, published a paper containing a systematic identification of the fossils discovered at the site since salvage operations began in the summer of 2000. Appearing in the November 26, 2002, issue of the *Annals of Carnegie Museum,* the paper lists sixteen different vertebrate animal groups, with mammals, fishes, birds, reptiles, and amphibians all represented by one or more species. Also included in the article are identifications of some invertebrates and plants. (A complete listing of these identifications, compiled from that article, appears as Appendix 2.)

Based on all of the identified fossils, especially those of the mammals, the deposit at Gray is thought to belong to the Hemphillian Land Mammal Age and dates to the Late Miocene–Early Pliocene time (4.5 to 5 million years ago). To arrive at these dates, researchers used the published identification of taxa by other researchers as a guide to help identify the newly discovered specimens. In time, radiometric-dating methods will be employed to further determine the age of the fossil bones. As research continues, the actual age of the fossil deposit may eventually be found to be slightly older or slightly younger than is presently thought.

Based on my own observations and on those of others involved in the discovery and salvage operations at Gray, I believe that the site was probably a water-filled sinkhole, bounded on at least one side by a limestone bluff. It was an environment that supported abundant plant and aquatic life in addition to surrounding terrestrial animals that lived in proximity

From left, Dr. Paul Parmalee, Dr. Steven Wallace, and Marta Adams gather at the spot where elephant bones were found.

These greenhouse structures, provided by the Tennessee Department of Transportation, give protection from the weather as excavation of the fossils continues at the site.

to the water. The description I offer in the prologue of this book represents my best guess as to what this prehistoric landscape may have looked like.

Quite possibly, the water that formed the small lake issued from a cavern as a spring, which was exposed by the sinkhole collapse. And it is likely that this pond environment lasted for a long period of time—the exact length is still unknown—before changes in the landscape and climate sealed its remnants in the earth until humans discovered them in May 2000.

Whatever the precise explanation for its origins and character, the Gray Fossil Site has proven to be a Tennessee treasure. It promises to enlighten scientists, students, and the general public for generations to come.

Appendix 1
Chronology

Spring 2000	Tennessee Department of Transportation contracts for a reconstruction project on State Route 75 in Washington County near Gray, Tennessee.
May 15, 2000	Lanny Eggers (TDOT construction project manager) has Freddie Holly contact Harry Moore (TDOT geotechnical manager in Knoxville) with news that soft black and gray clays have been encountered on S.R. 75 road project, hindering equipment; Holly asks Moore to visit project and recommend remedial action.
May 17, 2000	Larry Bolt (TDOT geologist from Knoxville) visits S.R. 75 construction project site to inspect clays.
May 25, 2000	Bolt informs Moore about his investigation of clays.
May 31, 2000	First fossil bones found at project by Bolt and Martin Kohl (Tennessee Department of Environment and Conservation); additional bone fragments found same day by Bob Price and Peter Lemiszki (both of TDEC).
Early June 2000	Martin Kohl gives some bone material to Paul Parmalee (University of Tennessee McClung Museum) and Walter Klippel (UT Department of Anthropology) for inspection; Parmalee and Klippel identify most of fossils as tapir bones.
June 8, 2000	Personnel from TDOT Geotechnical Engineering office in Knoxville visit Gray road construction project, finding vertebrate bones and teeth. TDOT staff present: David Barker, George Danker, Larry Bolt, and Harry Moore.
June 14, 2000	Bolt and Marta Adams (UT anthropology student) return to construction site to salvage additional fossil bones; Adams determines that bones are not human.

June 20, 2000	Moore and Vanessa Bateman (TDOT geologist) visit construction site and meet Larry Hathaway (TDOT inspector); Moore and Bateman unearth tapir skull and backbone.
June 24, 2000	Adams and Bateman visit site to salvage additional bones; find another tapir skeleton, including skull.
June 26, 2000	Moore and Don Byerly (UT geology professor) visit construction site. Moore finds jaw and big tooth; also finds section of elephant tusk.
June 27, 2000	Adams finds large elephant leg bone at road construction site; Rick Noseworthy (TDOT biologist) finds alligator skull. Moore advises TDOT personnel in Nashville of possible major fossil find.
June 28, 2000	Adams, Tony Underwood (Adams's husband), Bolt, Hathaway, and Moore unearth large section of pelvis bone of elephant; TDEC geologists visit site and find additional tapir bones.
July 6, 2000	Story of major fossil find breaks in newspapers in Knoxville and Tri-Cities (Bristol, Johnson City, and Kingsport); area of fossil discoveries closed to further construction.
July 7, 2000	Bolt, Mike Clark (UT geology professor), and Moore visit construction site to meet Nick Fielder, state archaeologist for Tennessee; television and newspaper reporters interview all four.
July 13, 2000	Organizational meeting of research committee to study fossil site; Fielder presides. Gray Fossil Site adopted as official name.
July 20, 2000	Moore visits Gray Fossil Site, along with Bob Hatcher and Chris Wisner (both of UT Geology Department) and Jeff Muncey (Tennessee Valley Authority), who conduct study on soft sediment deformation at the site. Two test pits dug by UT archaeology personnel and Fielder to determine extent of fossil deposit; tapir skeleton found six feet beneath surface.
July 25, 2000	Second meeting of Gray Fossil Site research team held at fossil site, with over twenty persons in attendance, including contractor; options discussed about what should happen to site.

August 2, 2000	Moore and Fielder meet at fossil site to discuss upcoming visit by Gov. Don Sundquist and his staff.
August 7, 2000	Governor Sundquist visits site for press conference, accompanied by Commissioner of Transportation Bruce Saltsman and Commissioner of Environment and Conservation Milton Hamilton. Fielder and Moore give governor tour of fossil site and display some discoveries.
August 16, 2000	Third meeting of research team held at fossil site with top administrators from TDOT in attendance. Options for site discussed; overwhelming majority of group favors moving road away from fossil site.
August 22, 2000	Hathaway and Moore meet at site to inspect progress of protection fence construction; find perfectly preserved tapir mandible at location where Governor Sundquist held press conference on August 7.
August 25, 2000	Parmalee receives information from Illinois State Museum that large tooth and jawbone found by Moore and second jaw and tooth found by Adams are probably from rhinoceros. Museum experts advise that materials be sent to Michael Voorhies at University of Nebraska, Lincoln.
September 12, 2000	Parmalee receives word from Michigan State University researchers that turtle remains found at site are from water turtles that lived in swampy waters or ponds, confirming Parmalee's belief that site fossils are probably older than "Ice Age" Pleistocene Epoch.
September 15, 2000	Governor Sundquist and Commissioner Saltsman visit Gray Fossil Site to announce that road construction will be relocated away from fossils to preserve them for research and education.
September 18, 2000	Parmalee and Klippel announce that fossil bones sent to University of Nebraska are indeed those of rhinoceros; this places age of Gray Fossil Site at 4.5 to 5 million years old or older.
May 2001	East Tennessee State University announces position open for vertebrate paleontologist, who will study Gray Fossil Site.
September 2001	ETSU hires Steven Wallace, Ph.D. graduate of University of Iowa, to study Gray Fossil Site.

October 9, 2001	TDOT makes and erects new sign for site.
December 2001	TDOT begins erection of four greenhouse structures at site to be used for field studies and fossil research.
March of 2002	TDOT begins construction of relocated S.R. 75 to bypass fossil deposit.
September 26, 2002	Governor Sundquist visits Gray Fossil Site and announces $8 million grant from Federal Highway Administration to construct museum, visitors' center, and educational facility at site.

Appendix 2
Animal and Plant Fossils at the Gray Fossil Site

The following identifications appeared in the article "A Late Miocene–Early Pliocene Population of Trachemys (Testudines: Emydidae) from East Tennessee," by Paul W. Parmalee, Walter E. Klippel, Peter A. Meylan, and J. Alan Holman. It was published in *Annals of Carnegie Museum* 71, no. 4 (Nov. 26, 2002).

Vertebrate Animals

Bear (cf. *Ursus* sp.)
Crocodilian (*Alligatoridae*)
Frog (*Ranidae*)
Mastodont (*Mammut* sp.)
Mustelid (*Mustelidae*)
Peccary (cf. *Catagonus* sp.)
Rhinoceros (*Teleoceras* sp.)
Salamander (*Caudata*)
Shrew (*Soricidae*)
Sloth (*Megalonyx* sp. or *Pliometanastes* sp.)
Small fish (*Osteichthyes*)
Small mouse (*Rodentia*)
Snake (cf. *Sistrurus* sp., cf. *Regina* sp.)
Tapir (*Tapirus*, cf. *T. polkensis*)
Turkey (*Meleagris* sp.)
Turtle (*Trachemys* sp.)

Invertebrate Animals

Pea clam (*Spheriidae*)
Snail (*Planorbidae: Helisoma* sp?)

Plants

Acorn (*Quercus* sp.)
Grape seed (*Vitis* sp.)
Hazel nut (*Corylus* sp.)
Hickory nut (*Carya* sp.)

The plant fossils also included remnants of leaves, tree trunks, and limbs; however, detailed studies of the macroflora have not been undertaken as of this writing.

Bibliography

Adams, Frank D. *The Birth and Development of the Geological Sciences.* New York: Dover Publications, 1954.

Bristol, Larry. "The Miocene: An Overview." *Friends of the Gray Fossil Site Newsletter* 1, no. 1 (2001): 5–7.

Farlow, J. O., J. A. Sunderman, J. J. Havens, A. L. Swinehart, J. A. Holman, D. L. Richards, N. G. Miller, R. A. Martin, R. M. Hunt Jr., G. W. Storrs, B.B. Curry, R. H. Fluegeman, M. R. Dawson, and M. E. T. Flint. "The Pipe Creek Sinkhole Biota, a Diverse Late Tertiary Continental Fossil Assemblage from Grant County, Indiana." *American Midland Naturalist* 145 (2001): 367–78.

Klepser, Harry. J. *Historical Geology Laboratory Manual.* Knoxville: Dept. of Geology, Univ. of Tennessee, 1967.

Klippel, Walter E., and Paul W. Parmalee. "Armadillos in North American Late Pleistocene Contexts." In *Contributions in Quaternary Vertebrate Paleontology: A Volume in Memorial to John E. Guilday,* edited by Hugh H. Genoways and Mary R. Dawson. Special Publication, no. 8. Pittsburgh: Carnegie Museum of Natural History, 1984.

Kohl, Martin. "Gray Fossil Site." *Newsletter of the Tennessee Division of Geology, Tennessee Department of Environment and Conservation* 13 and 14, no. 1 (2001): 8, 15.

———. "Personal Account of the Gray Fossil Site Discovery." *Newsletter of the Tennessee Division of Geology, Tennessee Department of Environment and Conservation* 13 and 14, no. 1 (2001): 16–17.

McDade, Arthur. *The Natural Arches of the Big South Fork.* Knoxville: Univ. of Tennessee Press, 2000.

Miller, Robert A. *The Geologic History of Tennessee.* Bulletin 74. Nashville: Tennessee Division of Geology, 1974.

Moore, Harry L. *A Geologic Trip across Tennessee by Interstate 40.* Knoxville: Univ. of Tennessee Press, 1994.

———. "Miocene Fossils Discovered on Tennessee DOT Road Project." In *Proceedings of the 52nd Highway Geology Symposium.* Cumberland, Md., May 16–18, 2001.

Moore, Harry, and Fred Brown. *Discovering October Roads: Fall Colors and Geology in Rural East Tennessee.* Knoxville: Univ. of Tennessee Press, 2001.

"Paleontology at the U.S. Geological Survey." *U.S. Geological Survey.* http://geology.er.usgs.gov/paleo/.

Parmalee, Paul W. "A Late Pleistocene Avifauna from Northwestern Alabama." In *Papers in Avian Paleontology,* edited by Kenneth Campbell. Science Series, no. 36. Los Angeles: Natural History Museum of Los Angeles County, 1992.

Parmalee, Paul. W., A. E. Bogan, and J. E. Guilday. "First Records of the Giant Beaver (Castoroides ohioensis) from Eastern Tennessee." *Journal of the Tennessee Academy of Science* 51, no. 3 (1976): 87–88.

Parmalee, Paul W., R. D. Oesch, and J. E. Güilday. *Pleistocene and Recent Vertebrate Faunas from Crankshaft Cave, Missouri.* Report of Investigation, no. 14. Springfield: Illinois State Museum, 1969.

Parmalee, Paul W., and R. D. Oesch. *Pleistocene and Recent Faunas from the Brynjulfson Caves, Missouri.* Report of Investigation, no. 25. Springfield: Illinois State Museum, 1972.

Parmalee, Paul W., Walter E. Klippel, Peter A. Meylan, and J. Alan Holman. "A Late Miocene–Early Pliocene Population of Trachemys (Testudines: Emydidae) from East Tennessee." *Annals of Carnegie Museum* 71, no. 4 (Nov. 26, 2002): 233–39.

Prothero, Donald R. "The Rise and Fall of the American Rhino." *Natural History,* August 1987, 26–32.

Prothero, Donald R., Claude Guerin, and Earl Manning. "The History of the Rhinocerotoidea." In *The Evolution of Perissodactyls,* edited by D. R. Prothero and R. M. Schoch. New York: Oxford Univ. Press, 1989.

Prothero, Donald R., and Robert M. Schoch. "Origin and Evolution of the Perissodactyla: Summary and Synthesis." In *The Evolution of Perissodactyls,* edited by D. R. Prothero and R. M. Schoch. New York: Oxford Univ. Press, 1989.

Spencer, Elizabeth H. *Basic Concepts of Historical Geology.* New York: Crowell, 1962.

Voorhies, Michael R. "A Miocene Rhinoceros Herd Buried in Volcanic Ash." In *National Geographic Society Research Reports 1978,* edited by Winfield Swanson. Washington, D.C.: National Geographic Society, 1985.

Wilson, Robert L. *Guide to the Geology along the Interstate Highways in Tennessee.* Report of Investigation, no. 39. Nashville: Tennessee Division of Geology, 1981.

Index

References to illustrations are printed in **boldface.**

The Bone Hunters was designed and typeset on a Macintosh computer system using QuarkXPress software. The body text is set in 11/14 Adobe Garamond and display type is set in Franklin Gothic. This book was designed and typeset by Cheryl Carrington and manufactured by Thomson-Shore, Inc.